INNOVATION POLICY

IRELAND

ORGANISATION FOR ECONOMIC CO-OPERATION AND DEVELOPMENT

Pursuant to article 1 of the Convention signed in Paris on 14th December, 1960, and which came into force on 30th September, 1961, the Organisation for Economic Co-operation and Development (OECD) shall promote policies designed:

  – to achieve the highest sustainable economic growth and employment and a rising standard of living in Member countries, while maintaining financial stability, and thus to contribute to the development of the world economy;
  – to contribute to sound economic expansion in Member as well as non-member countries in the process of economic development; and
  – to contribute to the expansion of world trade on a multilateral, non-discriminatory basis in accordance with international obligations.

The original Member countries of the OECD are Austria, Belgium, Canada, Denmark, France, the Federal Republic of Germany, Greece, Iceland, Ireland, Italy, Luxembourg, the Netherlands, Norway, Portugal, Spain, Sweden, Switzerland, Turkey, the United Kingdom and the United States. The following countries acceded subsequently through accession at the dates hereafter: Japan (28th April, 1964), Finland (28th January, 1969), Australia (7th June, 1971) and New Zealand (29th May, 1973).

The Socialist Federal Republic of Yugoslavia takes part in some of the work of the OECD (agreement of 28th October, 1961).

Publié en français sous le titre:

LA POLITIQUE D'INNOVATION
IRLANDE

This report is part of the OECD series of innovation policy reviews in individual Member countries.

These reviews have two purposes: first to enable the country concerned to appraise the performance of the institutions and mechanisms which govern or influence the various fields (scientific, technological and industrial, but also economic, educational and social) which contribute to its innovation capacity. The review provides an opportunity to assess the depth of social awareness of the need for, and the causes of, innovation.

Second, the review helps to enrich the pool of available knowledge on the content of innovation policies and their role as an instrument of government. In this way OECD countries can derive lessons which should help them to perfect their own policies. Similarly, through this improved knowledge of the resources deployed by Member countries, the reviews help to strengthen international co-operation.

A flexible approach is adopted in regard to the focus, the methodology and the presentation of these reviews. Although each review is centred mainly on the specific problems of the country under review, there is a common procedure for preparing and conducting them. The process of review consists of the following stages:

-- The preparation of a background report by the OECD Secretariat or by the country under review;

-- An information mission: a team of examiners visits the country under review and contacts those responsible for the various policy fields including senior officials, industrialists and academics. The examiners also visit a certain number of laboratories, universities and public institutions as well as industries. The aim of this second stage is to supplement the information provided by the background report and to enable the examiners to formulate what they deem to be the main problems raised by the implementation of the various elements constituting the innovation policy under review: this constitutes the examiners' report;

-- A review meeting, held as a special session of the Committee for Scientific and Technological Policy, at which representatives of the country under review answer questions put by the examiners and the delegates of Member countries.

The documents related to the review are then published under the responsibility of the Secretary-General of the OECD.

As regards Ireland, this publication includes only the examiners' report, to which are annexed an account of the review meeting and a background note on the Irish institutions concerned with science, technology and innovation.

The present report was written in response to a request made by the Irish Department of Industry and Commerce.

The information mission took place on 16th-23th March 1985, and the review meeting was held on 14th October 1985.

The examiners' team

The members of the examiners' team were:

-- Mr. Gerard J. Gogniat -- President, Groupement pour la Promotion du Capital-risque, Switzerland.

-- Mr. Morten Knudsen -- Managing Director, Technological Institute, Denmark.

-- Mr. Jean-Eric Aubert -- OECD Secretariat.

*

*　　　　*

The examiners and the Directorate for Science, Technology and Industry of the OECD wish to express their thanks to the Irish authorities and to numerous Irish personalities for their invaluable assistance in the preparation of this report. They wish however to express their particular gratitude to Mr. Cormac O'Connor, Assistant Secretary, Department of Industry and Commerce, Mr. Sean Aylward who ensured the co-ordination of the review and Mr. Dermot O'Doherty who prepared the background documentation.

Also available

INNOVATION POLICY – FRANCE (February 1987)
(92 86 06 1) ISBN 92-64-12884-0 296 pages   £16.00   US$32.00   F160.00   DM71.00

STI – SCIENCE TECHNOLOGY INDUSTRY REVIEW. No. 1/Autumn 1986
(January 1987)
(90 86 01 1) ISBN 92-64-12888-3 129 pages   £8.00   US$16.00   F80.00   DM35.00
1987 Subscription (No.2/SPRING & No.3/AUTUMN):   £15.00   US$30.00   F150.00   DM66.00

RECOMBINANT DNA SAFETY CONSIDERATIONS. Safety Considerations for
Industrial, Agricultural and Environmental Applications of Organisms Derived by
Recombinant DNA Techniques (September 1986)
(93 86 02 1) ISBN 92-64-12857-3 70 pages   £6.00   US$12.00   F60.00   DM27.00

REVIEWS OF NATIONAL SCIENCE AND TECHNOLOGY POLICY:
AUSTRALIA (August 1986)
(92 86 05 1) ISBN 92-64-12851-4 120 pages   £7.50   US$15.00   F75.00   DM33.00
PORTUGAL (June 1986)
(92 86 04 1) ISBN 92-64-12840-9 136 pages   £8.00   US$16.00   F80.00   DM35.00

THE SPACE INDUSTRY. Trade Related Issues (November 1985)
(70 85 03 1) ISBN 92-64-12772-0 92 pages   £7.00   US$14.00   F70.00   DM31.00

BIOTECHNOLOGY AND PATENT PROTECTION. An International Review by
F.K. Beier, R.S. Crespi and J. Straus (September 1985)
(93 85 05 1) ISBN 92-64-12757-7 134 pages   £8.00   US$16.00   F80.00   DM35.00

SCIENCE AND TECHNOLOGY POLICY OUTLOOK – 1985 (June 1985)
(92 85 03 1) ISBN 92-64-12738-0 98 pages   £5.50   US$11.00   F55.00   DM24.00

INDUSTRY AND UNIVERSITY. News Forms of Co-operation and Communication
(October 1984)
(82 84 04 1) ISBN 92-64-12607-4 70 pages   £3.50   US$7.00   F35.00   DM16.00

TABLE OF CONTENTS

SUMMARY

Introduction

The most critical economic problem facing Ireland is unemployment, which accounts for more than 17 per cent of the labour force. The government is actively addressing this in a variety of policies. The development of industry and services will be the main avenues for the creation of wealth and employment. In that context, it is essential for Ireland to have an aggressive innovation policy. This report addresses the many elements in such a policy and makes specific recommendations for action. In particular, the report stresses the need for a coherent approach, based on an overall vision for the future, a vision which should be actively developed and shared by the community as a whole.

The challenge

Some of the opportunities for new sustainable employment will come, as in the past, from overseas investment in Ireland. But there are already diminishing returns in employment terms from this generously-assisted investment, a fact that is fully recognised in the new industrial policy set forth in the Government's White Paper of 1984. The new policy incorporates:

-- A more selective approach to the use of industrial incentives;

-- A shift of government resources from support for fixed assets investment to technology acquisition and marketing development;

-- The increased integration of overseas investment in Ireland through greater linkage with indigenous industry and educational institutions;

-- The development of a risk capital market for indigenous industry;

-- Steps to improve the general business environment;

-- Measures to promote effective education, training and worker mobility; and

-- The development of a comprehensive industrial information framework to assess and review industrial policy.

This demonstrates a new emphasis on finding ways and means to unlock the potential of existing and new indigenous industry to meet the wealth and employment creation challenges of the 1980s and beyond. This must be based on an indigenous industry, competitive in international markets and based on high quality, high value added and high productivity. The new technologies offer many opportunities in this regard.

We are on the threshold of a new technological era. A very large array of technologies are in various stages of development: microelectronics in various forms -- computers, telecommunications, production automation, new industrial materials and biotechnology.

For small countries such as Ireland, the application and use of these technologies are critical and Ireland should seek to use them in a "labour-enhancing" way.

In terms of innovative orientation, Ireland is currently ranked in the bottom quartile of industrialised countries. The factors responsible for this appear to be:

-- Lack of specific technological skills and of a general awareness of new technologies in Irish firms;

-- The production of undifferentiated, low technology products sold at a low price, instead of products based on quality, new technology and exclusive design;

-- Exclusive orientation to relatively small, unsophisticated Irish and/or traditional British markets;

-- The absence of larger progressive indigenous firms which would provide a strong technological and industrial background for smaller companies;

-- Problems of communication and lack of acceptance of the need for technological change by management and workers;

-- Lack of opportunity for private individuals to make profit, partly because of heavy personal and capital gains taxes and, therefore, lack of development and venture capital; and

-- A general lack of understanding of innovation and entrepreneurship, and of a strong innovative culture within firms and in Irish society generally.

A range of policy measures is being put in place to address these issues. This report calls for an accelerated effort in many of these areas.

The vision

A policy for promoting technological innovation, conducted with deter-mination, is needed to set the country on a new path of development. Central to this policy is the formulation of a clear vision for the country as a whole. Ireland can sustain a vision for its future on the concept of a "model for development". This would be based on:

-- Massive and continuous development of its human resources;

-- The best use of new technologies for "fertilizing" local resources with ambitious goals for employment creation;

-- Creation of an entrepreneurial climate using cultural features geared to private and local initiatives, "plugged in" to international networks of knowledge and technology; and

-- The search for excellence and sense of quality in all areas of national life ranging from industrial production to public management.

Such a project would generate widespread discussion, help open minds and plant seeds for a powerful drive forward in economic development -- a process which involves society as a whole.

People need to be made aware of actual achievements by Irish entrepreneurs and communities; they need better media coverage of the economic and industrial development issues and they need to become involved in discussion of these issues. Individuals need to be stimulated and motivated by participation in concrete projects of national significance, e.g. a nation-wide programme of quality improvement. Individuals' responsiveness and commitment to a shared vision of development can be improved if they are involved in the development and funding of the local infrastructure appropriate to the new technological age.

The principles for government action

It would be desirable that the new vision include fairly ambitious targets focusing on the present unemployment problem. As indicated in Chapter I, it would be necessary by the year 2000 to:

-- Increase the labour force in industry and related services by 100 000, considering the employment problems facing the country; and

-- Double Ireland's share of the world market for industrial products.

Three points are fundamental in order to move towards these targets and to achieve further progress in boosting industrial development.

a)   Entrepreneurial initiative and the role of government

Up to now Irish government has played a large role in industrial life. The emphasis must now shift towards stimulating private entrepreneurial initiative.

Ireland cannot afford to increase total public expenditure. Private investment and initiative must be stimulated and more effective public provision must come from a restructuring of the institutional arrangements and a re-allocation of existing assistance.

b)   Decentralisation and flexibility

The overall development process should be understood as a "bottom-up" process based on local initiative. There should be a clear policy of decentralisation involving local authorities and communities in the development process to the maximum extent.

A modern society is characterised by a rapid rate of change and success for a nation depends on the ability to adapt by having flexibility in education, industrial policy and in institutions of government. We recommend that the guidelines and regulations for these institutions should aim at decentralising responsibility within Irish administration. They should allow freedom for public institutions to choose ways to solve problems while setting out clear objectives and criteria against which their work will be judged.

c)   Support investment in "brains" rather than in fixed assets

Modern development relies more and more on investment in "brains". Price is no longer the single dominant factor in international markets. Advanced firms in the high technology area invest in research and development, marketing, training and software as much as in buildings, equipment and machinery.

Ireland should, as a key element in its development strategy, move towards knowledge intensive industries while focusing on the factors most relevant to her needs and capabilities. The Irish innovation strategy should concentrate on: applying new technologies, developing market intelligence, continually upgrading the labour force, enriching management capabilities and emphasizing product and service quality improvement. A deliberate and sustained reorientation of resources towards education, research and development and information should be implemented.

Policy guidelines

The ability of Irish industry to meet the wealth and job creation challenges depends on an accelerated upgrading of its technological capacity and on a significantly increased rate of new enterprise formation.

a)   Support for new company formation

In line with the principles referred to above which should guide government action in creating the conditions needed for promoting innovation and setting up new firms, the experience in the Mid-West region of Ireland is a useful precedent. We believe that the Shannon experience should be studied with the aim of decentralising and co-ordinating locally the different types of support needed for talented entrepreneurs who have a bright idea but often lack the necessary marketing, financing and management expertise. The most effective form of delivery of the services currently offered to small industry is through a regionalised and co-ordinated service. It is recommended to continue the integrated support given to inventors and entrepreneurs through the Innovation Centre which has obtained promising results compared with similar institutions overseas. The comprehensive programmes of assistance for small scale start-ups should also be continued.

The "cult" of entrepreneurship needs to be actively promoted at several levels in Irish society: in basic education as much as in higher levels; within companies by changing the corporate climate in favour of "intra-preneurship"; within industry support schemes generally, by expanding schemes permitting entrepreneurs to explore opportunities abroad.

b)   Climate for investment and innovation

In both new and existing firms, undue reliance on grants to enhance the financial environment for entrepreneurs and innovators may be counter-productive for promoting an entrepreneurial spirit.

Private equity investment should gradually take the place of government grants. For this to happen, there need to be greater tax incentives for private savings in order to encourage future business creation and more generous tax treatment of capital gains in order to give greater recognition to the risks involved in industrial investment.

The development of venture capital sources and structures has elsewhere been an important tool for supporting innovative ventures, complementing the traditional sources of finance. This avenue is underexploited in the Irish context. A further reduction of the long term capital gains tax to no more than 20-25 per cent or preferably a total exemption, with an assurance of adequate longevity for this incentive would be desirable to boost the return on such investment and make it at least as attractive as other less immediately "productive" avenues.

Venture capital funds which involve patient money and hands-on manage-ment expertise need to be allowed a reasonable operating margin to ensure the hiring of competent management -- capable of and willing to take and manage above-average risks.

c)   Upgrading of technological capacity

A doubling of investment by industry in "brain-power" or skill-based assets will be necessary in the medium term and the implementation of a deliberate and sustained reorientation of the industrial grant support system to encourage this is recommended.

The additional investment in skills should go hand in hand with the building up of the indigenous technical culture supplemented by effective transfer of technology from overseas.

An integrated approach by industry itself, by technological services institutes and by government, will be required. Efforts made in the past to raise the technological level of the Irish agriculture sector might provide a model for such an integrated approach. The technological resource base of foreign industries established in Ireland should be tapped more effectively for a contribution to improving the technical capacity and skills of indigenous industry. A commitment by foreign industry to maximise its contri-bution in this way might, for instance, be made a precondition for any new grant aid.

d)   Technological support services

The development and rationalisation of a nation-wide network of technological services related to the needs of industry requires additional resources, stronger commitment by industry itself and commitment at the local or community level.   In Chapter V we recommend possible ways in which additional resources can be raised, the activities which should be funded from these new resources and the ways in which industry should be helped with this assistance to strengthen its own demand for further technological effort.   A system of the kind proposed would provide a suitable model for ensuring an appropriate balance in research institute funding between guaranteed government funding for the minimal infrastructure in personnel and equipment needed to satisfy properly the demands put on them, and discretionary funding from contract fees for serving local industry or from international contracts.

e)   Higher education-industry linkage

Indigenous Irish industry is using the eminent resource of highly skilled manpower in third level (sometimes referred to as "tertiary" level) institutions to only a very limited degree.   The steps already in train to develop this resource and to forge fruitful co-operation with industry should be further expanded to the mutual advantage of both parties.

Industry should become more involved in influencing third level education, and colleges should more actively market their expertise and facilities to industry.   In addition to the undertaking of additional research for industry, which would be part-funded from the new sources indicated above, we recommend certain institutional and other changes in the operation of these third level institutions which are necessary for the expansion and development of education-industry linkage.

f)   Education and training for innovation

If Irish industry is going to compete by means of technology, quality and good management, adjustments to the skill and qualification level of the workforce will be required.   At present Ireland has only one third of the number of engineers and technicians of industrialised countries.   In modern high technology industries over 50 per cent of the workforce have technician or higher grade skills.   If Ireland is to build her hopes on such industries the potential demand for skilled staff will increase significantly.   The output of engineers and other highly qualified graduates from third level institutions should, in our view, be rapidly accelerated especially in areas such as electronics, data processing, production engineering, materials science, chemistry, biotechnology, product development, strategic planning, economics and marketing.

The increase in the college intake of engineering and technology students alone should be of the order of 25 per cent per annum over the next 5 years.

There should also be expansion of the numbers and development of the courses in the colleges meeting the demand for technicians and skilled workers.   Training and re-education facilities for the existing workforce must also be expanded to enable them to master the new technologies and to meet other future qualification demands.

The responsiveness of the educational system at all levels to the future needs of a society based upon technology and international trade will be improved if there is regular and joint review of the curricula and qualifications by the institutions which provide the education and by the industry which ultimately creates the demand for particular skills.

In this respect the establishment of the Curriculum and Examinations Board as part of the process of reforming the second level curricula is seen as a welcome development.

Finally employers' associations and trade unions have an important role to play in education and training for innovation. Experiences from other countries, e.g. the Scandinavian countries, might provide an inspiration for similar developments in the Irish context.

I. INTRODUCTION: A VISION FOR IRELAND

Ireland has successfully undertaken, over the last few decades, a major effort to build up an industrial infrastructure and to create jobs in the manufacturing sector, particularly by attracting foreign firms through generous financial support and tax incentives for capital investment. This policy has reached obvious limits today, as shown by various signs -- increasing unemployment and the increasing cost of job creation.

A new leap forward will have to be largely based on the development of indigenous industry (notwithstanding the continuing importance of the foreign owned companies within the economy). More than six out of every ten workers in the manufacturing sector in Ireland are employed by indigenous industry. This is a resource which must be both conserved and used as a basis for future wealth and employment creation.

This sector is obviously in difficulties. To strengthen it as a source of wealth and jobs, it needs massive technological upgrading and to be oriented towards international markets. We believe this is possible.

A policy for promoting technological innovation, conducted with determination, is needed to set the country on this new path of development. Central to this new policy is the formulation of a clear vision for the country as a whole. The importance of such a vision was well demonstrated in the 1950s by the famous document on "Economic Expansion" (1). It set out targets for economic achievement which, though thought by some as beyond Ireland's grasp at the time, were met to the full.

*

*   *

Technological innovation sometimes occurs through major projects such as space exploration, satellite telecommunications, etc., but more often manifests itself in day to day applications like well-designed clothes, better-tasting food, more durable cars, services delivered more speedily, or safer working conditions.

The end of this century is characterised by a technological revolution based on the development and diffusion of a series of key technologies such as microelectronics, new materials, biotechnology. Innovation will mainly result from the appropriate application of these new technologies to renew existing products and services, improve manufacturing processes and exploit natural resources.

17

Ireland has good assets for innovation. Paramount among those are: the youthfulness of her population, the inner dynamism of Irish people, the capability to mobilise energy in local communities, and openness on the international scene through active participation in the European Community and long-established links with North America.

Several examples of successful achievements by Irish firms and entrepreneurs ranging from high technology areas (e.g. in remote sensing by satellite) to more traditional ones (e.g. in liqueurs) could be quoted to illustrate the existence of an innovative potential. However, those cases are still too infrequent in relation to the considerable challenge that the country has to face. So there is a need both to expand this potential and to release it.

*

*   *

A new vision should encompass precise targets for the country as a whole. To solve the present and future unemployment problems, it would be necessary by the year 2000 to have:

-- Increased the labour force by about 100 000 (up to 320 000) in industry and related services (e.g. software houses, engineering);

-- Doubled Ireland's share of the world market for industrial products, which is currently less than one per cent.

These goals may appear over-ambitious but they must be kept in mind if the unemployment and economic growth problems facing the country have to be solved.

In moving towards these goals it is recommended in particular that:

-- In the next 3-5 years industry double in volume its investment in "brain power" (training, research and development, design, marketing research, computer software...);

-- The government take immediate steps to support these investments in "brain" as much as investments in fixed assets (plant, building, machinery, hardware...);

-- The intake of engineering and technology students be increased by 25 per cent per annum over the next 5 years; and

-- Boys and girls in secondary school be exposed wherever possible to practical work experience and more particularly be familiarised with information technology including the application of computers.

Meanwhile the government should take a series of measures to improve the climate for innovation and entrepreneurship:

-- In favouring by all available means (notably fiscal incentives) equity investment in industry by private individuals. Equity investment as part of the capital in manufacturing industry is remarkably low -- 10-15 per cent for new manufacturing and technology-based companies -- and should be doubled within the next 10 years;

-- In stabilizing and, if possible, reducing the overall public expenditure which leads to an increasing burden of government debt and heavy taxation; and

-- In devolving responsibility and instituting decentralisation within the public sector, especially in those institutions supporting technological development.

*

* *

The rationale for these proposals as well as other important measures is developed in the following chapters, which attempt to chart a comprehensive innovation policy to meet the challenges that Ireland is facing in this new technological age.

Chapter II makes more explicit the innovation challenge that Ireland has to face. Chapter III sets out the fundamental principles and reorientations that should guide government action if further progress is to be made.

The three following Chapters address what we feel to be the key components of the innovation policy which should be pursued in Ireland. Specific policy suggestions are made in relation to the support to be provided to entrepreneurs and innovators (Chapter IV), to methods for raising the technological level of industry (Chapter V) and to the need for intensifying the education effort (Chapter VI).

Chapter VII formulates some indications on how to sustain the new vision presented in this introduction and make it credible and concrete.

II. THE INNOVATION CHALLENGE

Past and present performance

Industrial development policy has been a major concern of Irish policy makers for more than 50 years. Following the achievement of independence by the Irish Free State in 1921, governmental efforts to promote the development of the country's very weak industrial base were limited mainly, at first, to infrastructure development projects.

In the early 1930s, however, strong protection against imports from more advanced economies was introduced on a wide scale in order to foster development of new "infant industries" and promote greater economic self-sufficiency. Industrial employment grew rapidly under protectionist policies until the early 1950s. This growth, however, took the form of import-substitution and few industries developed a capability to compete in export markets. Since continuing growth depended on large scale imports of the many materials, components, capital goods and fuel still not available from a small domestic economic base, a chronic balance of payments crisis emerged in the 1950s, which virtually halted further growth during most of that decade and led to a fall in industrial employment and a substantial increase in emigration.

In this context of disillusionment with protectionist policies and promotion of self-sufficiency, a more "outward looking" approach was adopted in the 1950s. Foreign capital was sought for investment in export industries and many export-orientated subsidiaries of foreign companies were established in Ireland (with the aid of generous taxation and financial incentives) throughout the 1960s and 1970s, leading to quite rapid growth of industrial employment and significant gains in export shares. Ireland's protectionist barriers also began to be dismantled ; full free trade with the United Kingdom gradually came into operation, starting in the mid-1960s, and Ireland's accession to EEC membership in 1973 opened the home market to free trade with a wider group of European economies.

Attracting foreign capital for investment in export oriented industries has led to rapid growth of production and gains in export shares. Thus Ireland keeps a leading position in relative production growth inside the EEC, i.e. over 40 per cent in the five year-period 1978-83, a relative growth comparable to that of Japan, and three times the average relative growth of production in the EEC.

Seen from that point the policy has been a success so far. Ireland is on the move.

However, relative employment performance in manufacturing has not been as robust as output or exports, particularly since 1980. Manufacturing employment grew by 9 per cent in the 1975-80 period, while average employment in the EEC declined by 4 per cent. But since 1980, Irish manufacturing industry has shed labour rapidly, losing 13 per cent of total employment, very similar to losses experienced elsewhere in the EEC. The current rate of unemployment is 17 per cent, with all the related problems that can be expected. Migration outflows, stopped and even reversed since the end of the 1950s, are now slightly positive. Of particular concern is the fact that this migration involves well-educated and talented people such as managers, engineers or designers who cannot find job opportunities in Ireland.

Thus the industrial policy so far pursued has met obvious limits. A recent OECD Economic Survey of Ireland (2) presents the main issues as follows:

-- "The total cost of financial and fiscal incentives has been high. The total direct and indirect costs of this policy may have amounted to Ir£800 millions, around 6 per cent of GNP. The cost of grants alone per sustained job averaged about five times average annual earnings in the early 1980s;

-- A dual industrial economy has emerged, with a poor performance of traditional indigenous industries, output gains and export expansion being largely concentrated in foreign owned high-technology sectors; inter-industry linkages between foreign owned and domestic sectors have been weak, limiting Irish value added in foreign owned operations to direct wages and salaries;

-- Reliance on foreign investment to create a modern industrial structure has made the Irish economy vulnerable to slowdown in new foreign investments. It is unlikely that Ireland will be able to rely on rapid increases in foreign investment to boost industrial performance to the same extent as in the past."

It is therefore clear that the future development of Ireland should be based much more on indigenous industry. The indigenous industry should become competitive on the world market based upon free market conditions, i.e. high quality, high value added and high productivity instead of low price and governmental grants.

The "Telesis report" (3) confirmed that Irish indigenous firms contributed relatively little to expansion in the 1970s, have developed few new export activities, and have remained largely confined to sheltered, non-traded and traditional, technically undemanding, relatively labour-intensive industries. One of its key recommendations was that industrial policy should aim to increase the value added per worker embodied in the goods and services produced in internationally traded industries. This was to be partly achieved by improving productivity in existing activities, but more fundamentally by continually shifting the composition of Irish industry out of easily-entered, low-wage activities into different activities which can sustain higher incomes.

Table 1

EMPLOYMENT IN INDIGENOUS AND FOREIGN MANUFACTURING INDUSTRY:
1973, 1980 AND 1983

	1973	1980	1983 (Oct.)
Indigenous	158 241	163 483	137 030
Foreign	59 051	80 430	79 390
Total	217 292	243 913	216 420

Table 2

EMPLOYMENT IN INDIGENOUS INDUSTRY -- 1983

No. of employees	No. of companies	Total employment
500+	21	16 930
200-500	66	19 850
100-200	169	23 430
50-100	342	23 910
3-50	4 229	56 070
Subtotal	4 827	140 190
Less than 3	939	1 480
Total	5 766	141 670

Table 3

EMPLOYMENT IN INDIGENOUS INDUSTRY BY SECTOR -- 1983

Food	38 990
Drink and tobacco	6 400
Textiles	6 920
Clothing and footwear	12 310
Paper and print	12 450
Wood and furniture	11 000
Non-metallic minerals	12 500
Chemicals	4 750
Metals and engineering	27 360
Other manufacturing	4 350
Non-manufacturing	4 640
Total	141 670

Source: IDA 1984 annual report.

In spite of the problems faced by indigenous firms in adjusting to new technological and market conditions they still provide more than 60 per cent of total industrial employment, covering a wide range of individual sectors (see Tables). They are thus a resource which must be both conserved and used as the basis for future wealth and employment creation.

Very constructive initiatives have been already taken by the government as illustrated in the 1984 White Paper on Industrial Policy (4), the National Plan Building on Realities, 1985-87 (5) and other publications related to the theme. This report argues for an accentuation of these initiatives and offers suggestions for making them successful.

It is important at the outset to identify more clearly what the development objectives should be in terms of job creation.

Young Ireland and the need for jobs

The population of the Republic has been growing during the last decade with a birth rate as high as 1.5 per cent per year. Fifty per cent of the population is younger than 25 years. It is generally expected that this birth rate, which is the highest in the EEC, will decrease.

The White Paper indicates a prospective growth in the active labour force of 17 000 per year. Further, it is stated that the Government believes that manufacturing output can be doubled over the next ten years and that this doubling of output can lead to an overall growth of 3 000-6 000 jobs a year in manufacturing. This is ambitious but not enough.

If the very large unemployment problems are to be solved, it has to be recognised that, in addition to creating jobs for about 200 000 who now are unemployed, it is also necessary annually to find new jobs for a high percentage of the 60-70 000 newcomers to the labour force, who are the result of the birth rate, as practically all boys and girls want to have a paid job when they finish their education. Further, some women, who earlier have been working exclusively in the household, wish to get paid work. These figures indicate that there is a need to create much more than 17 000 jobs a year; probably 2-3 times that amount, if the unemployment problems are to be solved before the year 2000, without a new wave of emigration.

These figures are based on our rough estimates. They suggest that the White Paper figures may be strongly underestimated and therefore it is recommended that the forecast be revised. If the new forecast coincides with the conclusions above, 30-40 000 additional jobs are needed each year if unemployment problems are to be solved before the year 2000, without creating a new wave of emigration. And this has to be done in the European Community, where other countries are also suffering from serious unemployment problems.

In 1983 the total employment in manufacturing was about 220 000, i.e. 20 per cent of total employment. If this share is to be constant with a total annual increase in the number of jobs of about 30-40 000, it is necessary to create 6-8 000 jobs annually in manufacturing. If this share -- as a very attractive goal -- was increased to 25 per cent, the average net increase in jobs in industry has to be about 25 000 during the next 15 years, and the total employment in manufacturing in the year 2000 would be about 400 000, i.e. a 100 per cent increase.

Table 17. **Structural change in manufacturing industry**

	Industrial production		Employment[1]			Foreign employment			Domestic employment	Foreign new fixed asset shares[2] (%)	
	% change					% share					
	$\frac{1983}{1973}$	$\frac{1984}{1983}$	1973	1983	$\frac{1983}{1973}$ % change	1973	1983	$\frac{1983}{1973}$ % change	$\frac{1983}{1973}$ % change	1975-1977	1981-1983
Total manufacturing	44.9	13.3	217 292	216 420	-0.4	27	37	36.6	-10.8	48	48
of which:											
Chemicals	160.6	23.9	11 253	13 610	20.9	52	65	50.2	-11.3	57	55
Metals and engineering	88.4	27.0	42 513	64 970	52.8	47	58	90.4	20.2	59	69
Food	38.4	4.3	46 856	45 560	-2.8	15	14	9.8	-4.6	23	11
Drink and tobacco	20.4	1.4	10 790	10 230	-5.2	39	37	-9.5	-2.5		
Textile industry	-15.6	0.8	23 003	11 830	-48.6	23	42	-8.5	-60.8	66	31
Clothing, footwear	-28.4	-0.9	24 769	18 850	-23.9	21	35	24.0	-36.8		
Timber, wooden furniture	-14.2	-4.5	10 771	11 750	9.1	7	6	-7.5	10.3	11	78
Paper and printing	-7.6	-1.6	14 503	14 100	-2.8	12	12	-7.7	-2.1	0.4	24
Non-metallic production	11.8	5.1	17 282	15 190	47.7	50	18	1.4	93.0	54	30
Miscellaneous	53.1	-1.6		10 330			58				

1. Employment based on IDA surveys.
2. Investment shares based on IDA-approved investment under new industry scheme.
Sources: IDA, *Annual reports* (various issues); CSO, *Industrial Production Index* (various issues).

This might be far too optimistic, but a goal of having in the year 2000 an increase in the labour force of about 100 000 up to 320 000 in industry and related services (e.g. software houses, engineering) should be realistic, if the competitiveness of Irish industry is based upon talented use of technology, efficient management and internationally-oriented marketing.

What does this mean in relation to investment? In general terms an investment in modern industry amounts to a minimum of Ir£20 000 per job. Creating 100 000 new jobs therefore means a need for capital investment in the order of Ir£2 billion totally, or about Ir£130 million a year. This figure is high but should not be a major barrier, if the right climate for industrial investment is established.

A crucial question is whether there will be a fair chance for Irish industry to gain market share in competition with other countries which are trying to increase their share. We take the optimistic view that Ireland has a reasonable chance to double its less than 1 per cent share of world market for industrial products during the next 10-15 years, taking into consideration that a great proportion of the labour force will be young and well educated and motivated to make an efficient contribution to innovation. Moreover, the fact that the industrialisation process in Ireland is still at an early stage should be considered as a positive factor -- there are no "industrial cemeteries".

Taking advantage of the new technologies

We are on the threshold of a new technological era. A very large array of technologies are in various stages of development: microelectronics in various forms, computers, telecommunications, production automation, industrial materials, biotechnology.

This development is primarily led by companies in the US and Japan, while Western Europe, and especially small countries, like Ireland, are lagging behind. But for small countries what matters is more the application and use rather than the development of new technologies.

These new technologies are diffusing across the existing sectors, renewing myriads of products and processes and creating new ones. Just to take a few well-known examples, microprocessors are to be found in all sorts of objects -- cars, toys, etc. New glues, based on polymer chemistry, very cheap and easy to use, transform traditional industries such as shoe-making and furniture. Thus, manufacturing industry, and more generally society as a whole, should develop its receptiveness to, and make the best use of, this new technological wave.

During the two decades from about 1950 to 1970 the main consequence of technological development was the mechanisation, and, to a certain extent, the automation of production. The pre-eminent purpose of new technology was to save labour. Machines and plants were made still bigger and more complicated, and there was an obvious trend to concentrate production in ever larger units. The continued mechanisation/automation has made a great number of manual processes superfluous or has changed the job content radically. This development is likely to continue, but the leap in development, caused especially by information technology (IT), is going to mark a new era.

For innumerable fields of application, IT opens up the prospects of extensive automation and (of vital interest to a society dominated by small and medium-sized firms) a high degree of flexibility, making even short serial production profitable, because changing of tools and performing of other operations may be automated and directed by computers. It is gaining in all branches of production and in our daily life as well, i.e. not only in capital equipment, but also in consumer products, office work, construction, health services, professional and public information, etc.

IT will bring about considerable productivity improvements -- and will require further changes in job content and qualifications needed at all staff levels.

It is a strong tool, to be used in starting a development which opens up opportunities for a unique improvement in production and productivity, work environment, and information: a tool which facilitates decentralisation and the enlargement of the individual's influence on his/her work and daily life.

Countries which lag behind in the introduction of new technology risk much greater losses of jobs than those arising from the early use of the technology. In some advanced countries, the net result is a positive contribution to employment by technology, largely through its effects on employment in services. In Sweden for instance, recent data indicate an increase of the labour/output ratio and a relative decline of the capital/output ratio -- just the reverse of the trend during the last few decades. In Denmark, industrial employment has increased from 350 000 to 375 000 between 1983 and 1985. These examples make more credible the goals for job creation stated above.

Barriers to innovation in Irish industry

The latest European Management Forum report (6) on international industrial competitiveness of 22 Member countries of the OECD ranked Ireland 17th in terms of overall competitiveness. One of the aspects surveyed in this report was the innovative forward orientation of the Member States.

In arriving at the overall innovative forward orientation, 29 criteria were evaluated. Ireland was ranked in the bottom quartile in the following categories:

-- Strategies towards the future (firms planning effectively for the long term);

-- Innovativeness in the firms (both in production techniques and in product lines);

-- Corporate readiness to exploit inventions;

-- National creativity (patents per 100 000 residents);

-- Expenditure on R&D as per cent of GDP: the figure for Ireland is 0.7 per cent.

Turning to the manufacturing sector the R&D picture is summarised in Table 5.

Table 5

R&D PROFILE OF MANUFACTURING FIRMS IN IRELAND: 1982

	Foreign owned	Irish owned	Total
Number of plants	811	4 652	5 463
Number performing R&D	137	231	368
Percentage of companies performing R&D	17	5	6.7
Number of companies with formal R&D departments	82	74	156
R&D expenditure (Ir£m)	22.3	15.6	37.9
Total employment	82 000	147 000	229 000
R&D expenditure/employee (Ir£)	272	106	166
R&D expenditure/net output (%)			1.1

Source: National Board for Science and Technology.

In 1982 6.7 per cent of all manufacturing companies were doing R&D. However among foreign owned firms the percentage was 17 per cent and 5 per cent among indigenous firms. Over 10 per cent of the foreign owned firms had formal R&D departments compared to 1.6 per cent of the Irish owned companies. R&D expenditure per employee was two and a half times higher in foreign than in Irish firms.

It is often stated that most Irish firms are very small by European standards and therefore they are not geared to make the necessary investment in R&D and international marketing. This may be a fact to be considered when comparing with some other countries with industrial structures which differ essentially from Ireland, e.g. the USA, France and Germany. But compared with, for example, Denmark, which is characterised by a similar industrial structure regarding company size, i.e. dominated by small and medium-sized enterprises, it is not a factor which should be deemed to be an insuperable barrier to innovation.

Certainly the size of a company influences the ability to invest significant funds in R&D and marketing. But, as known from Denmark as well as from the USA and other industrialised countries, many small and medium-sized firms can, nevertheless, be very innovative and flexible, even if they do not spend large sums on R&D or market research. Given that Irish firms do not in general act in this way, although there are some notable exceptions, one has to draw the conclusion that a combination of problems exists within firms themselves, in the structure of their industries and in their general economic and social environment. Among these factors, the most critical are probably the following:

-- Lack of specific technological skills and little awareness of new technologies in Irish indigenous industry (SME);

-- The production of undifferentiated, low technology products sold at a low price, instead of products based upon quality, new technology and exclusive design;

-- Exclusive orientation to the relatively small, unsophisticated Irish and/or traditional British market;

-- The absence of larger progressive indigenous firms which would provide a strong technological and industrial background for smaller companies;

-- Problems of communication and lack of acceptance of the need for technological change by management and workers;

-- Lack of opportunity for private persons to make profit, partly because of heavy personal and capital gains taxes, and therefore lack of venture capital; and

-- A general lack of understanding of innovation and entrepreneurship, of a strong innovative culture within firms and in Irish society generally.

If indigenous Irish industry is to reverse its current decline, to innovate and gain new markets, barriers like these have to be eliminated. If they are not removed, or at least reduced, there is a serious risk that a large part of the indigenous Irish industry will continue to lose competitiveness on international markets and retain its share of the home market only in those areas which have natural protection against imports, i.e. non-traded goods based on local resources. There is therefore a need for a national innovation strategy aimed at upgrading and repositioning Irish indigenous firms. Properly implemented such a strategy should be successful.

Recent empirical studies support this conclusion. The recent study, "Innovation in Established Irish Industry" (7) which follows up a survey carried out ten years ago, shows that the situation is not totally bleak. "If the task facing the companies is substantial there are, however, grounds for optimism. These companies have survived in difficult circumstances and they want to change. They have a history of some success in innovation : by comparison with ten years ago they have increased their rate of innovation, shifted it more towards product innovation, and enjoyed very high success rates with their attempts -- especially the more conservative ones. Furthermore, even if they are short of qualified staff, they have in many instances the nucleus of a development team with a history of successful projects. Most significant of all, however, they have the example of firms newly established in their own industrial sectors. These newer firms achieve a consistently high rate of successful product and process innovations, often directed at export markets."

These findings, based on recent empirical investigations, both reinforce the need for, and assure the probability of success of, a national innovation strategy aimed at upgrading and repositioning Irish indigenous firms.

People: the major resource

If a number of barriers to innovation exist in Ireland, the country possesses also good assets for developing its innovative capabilities. We would like to stress here the human resources.

A young population is an important asset. New generations provide dynamism and change the general "state of mind". Youth creates difficult challenges in terms of job creation but this could be converted to a tremendous force for progress if good education is provided and opportunities for self expression through entrepreneurship and innovation are given.

Ireland has also a long history of emigration. The large number of persons of Irish descent living in the United States and other countries, e.g. Australia and Canada (40 million), represents a considerable reservoir of goodwill for development of international contacts for Ireland with those countries.

Irish people have also established close links with foreign organisations -- multinational firms, universities, international institutions -- where they have learnt a great deal as managers, teachers, civil servants. These experiences and networks should be used more than they are at present.

Finally there is an important point related to Irish culture. The recent US bestseller in management science In Search of Excellence (8) identifies the existence of a "corporate culture" as a basic ingredient to successful firms. This culture provides an identity in which the firm finds its driving force. The same concept applies to individual countries. What are the features of Irish culture and identity?

The sense of co-operation in small communities for survival and development might be one of the most favourable features of Irish culture.

The practice of neighbours helping each other with harvesting is as old as time itself and is still practised. The Irish co-operative movement in agriculture grew from this tradition and we see it also at work today in community-sponsored social projects throughout Ireland, like the SHARE movement in Cork which has built many houses for old people.

In relation to business and industry, we also noted a growing number of community-based initiatives taken at the level of small towns or suburbs. For instance, in the suburbs of Dublin, private individuals joined by the banks and stimulated by local authorities, have recently funded and established an "incubator space" offering premises and other facilities to small business start-ups, e.g. designers, craft-workers, etc.

Building on such a socio-cultural asset would certainly be a sound approach. This may offer a middle way between total State support and a totally laissez-faire situation. The recommendations which follow take this feature into consideration.

III. PRINCIPLES FOR GOVERNMENT ACTION

For historical and structural reasons, the government has played a major role in the development of Ireland. Its role appears particularly significant in the industrial field through the importance acquired by several public institutions, notably the Industrial Development Authority. The success achieved in boosting industrial development and in sustaining the overall economic take off is unquestionable. But there is now a need for radical reorientation if any further progress is to be achieved. Three fundamental points will be discussed below:

-- The need to increase entrepreneurial initiative;

-- The need to promote decentralisation and flexibility in administration; and

-- The need to support investment in brains rather than investment in fixed assets.

Entrepreneurial initiative and the role of government

The public sector absorbs more than 60 per cent of GNP. Such a high share of the public sector in national life brings with it a large public debt and a heavy taxation burden on the population. Moreover, it tends to induce a mentality which places primary responsibility for problem solving on the State rather than on personal initiative, and to reinforce a tendency to look to the State for support.

The natural generosity of Irish people finds remarkable applications today, for instance when they help massively and quickly in solving problems of famine in the Third World, through financial and other support.

However, with regard to the development of Ireland herself, this natural generosity tends to bring with it the idea that everything can be solved or promoted by grant assistance; there is a clear need to find means to reorient this generosity towards a more efficient self-assistance approach.

Insufficient attention appears to be given in the current national economic plan 1985-1987 to the desirability of a greater role for entrepreneurial initiative in spite of the magnitude of the goals decided upon. We are particularly concerned with the effect of probable low individual motivation in the context of the role expected from innovation for the medium term development of Irish indigenous industry. In the long run it is probable that more attractive general conditions, e.g. lower individual taxation and lower capital costs, as well as incentives for equity and venture capital, will be as important for economic development as direct supports for industry.

In this sense, we would suggest an acceleration of the process of gradual reduction, not only stabilization, of public expenditures permitting further reductions in general taxation, particularly the marginal rate; at the same time measures should be taken to raise the propensity for private savings and facilitate their flow into industrial investment: a change which should build upon the idea of self-reliance in preference to the belief in development primarily through State grants.

Moreover, new initiatives should not increase total public spending but should include restructuring and re-allocation of existing sources of assistance, agencies, institutions, etc. There has been a tendency in Ireland to create new tools or agencies rather than looking at the less expensive possibilities of changing existing structures and attitudes. As indicated in the White Paper on Industrial Policy, this attitude may now be changing.

For example, when new themes show up, like microelectronics, it is generally worthwhile considering if an existent institution may be, after some changes, the centre for a new activity -- instead of creating a new organisation.

These considerations may also be relevant to the issue of the State commercial companies. The State is, and will remain, heavily involved in the commercial life of this nation. State national companies must be as competitive, efficient and enterprising as privately owned companies. They have also an important role to play in constituting the industrial and technological background that small subcontracting firms need. However it is not advisable that State companies or subsidiaries be created to undertake activities which should be normally developed by the private sector. The information technology sector is a particular case in point where the private sector in many instances is well capable of providing back up services to manufacturing industry.

Decentralisation and flexibility

The development process should be understood as a bottom-up process based on local initiative and not as a top-down State controlled process. In relation to this, two issues merit particular consideration in Irish society: decentralisation and flexibility.

Ireland is a highly centralised State. More than 30 per cent of the population lives in the capital and its area, a situation almost unique among OECD countries. A centralised bureaucracy tends to concentrate the intellectual potential of the country and the decision power in conservative institutions. As a natural consequence, public institutions are often more interested in expanding or defending their territories than in real and efficient public service (and thus tend to constitute a kind of new feudalism).

The experience in the Mid-West region constitutes a useful precedent for the successful conduct of the decentralisation process and for the creation of the conditions needed for promoting innovation and setting up new firms. Paramount among these conditions have been a good infrastructure, including Shannon Airport, the establishment of strong education facilities, including the National Institute of Higher Education of Limerick, and the Shannon Free Airport Development Company SFADCo. Shannon is an exception in Irish administrative history; it has operated with a large degree of autonomy, great flexibility of tools (taking action through grants but also in training, research and technical assistance), and a good knowledge of the "milieu" through a network of field officers in close contacts with firms.

The cost of the Shannon experience has been high and some of its initiatives could be questioned. However, it is undeniable that a new climate has been created in the region, characterised by a remarkable entrepreneurial dynamism. This is manifest not only in technology and industry in a narrow sense, but also in renovation of towns, adaptation of Shannon Airport services, and development of cultural activities.

According to SFADCo, employment in small manufacturing firms (less than 50 employees) has increased by fifty per cent in five years (1979-84), and the number of new firm start-ups has doubled during the same period of time. These are remarkable achievements compared to similar experiences in other OECD countries, although the Shannon region is still very heavily dependent on foreign owned branch plants for industrial employment.

Geographical concentration of intellectual potential would appear a prerequisite for modern development as illustrated by experiences such as California's Silicon Valley and Japan's Technopolis. This points to the need for adequate institutional arrangements and greater initiative for regions and counties.

Maybe Ireland is not ready to commit herself to a broad decentral-isation process along the lines of the Shannon-Limerick area. But there should be a clear policy to reverse the trends towards centralisation and to involve local authorities and communities as much as possible in the develop-ment process, by adequate policy measures. Specific suggestions will be made later in this report.

A modern society is characterised by a rapid rate of change. Much of this change is caused by new technology and new market situations. The success and failure of a nation depends to a high degree on its ability to cope with changes in vital factors like these. This imposes great demands on the flexibility of a nation, i.e. flexibility in education, industrial policy and in established institutions. Flexibility might be the most critical factor in Irish society in general, and in public service in particular. Nowhere is this flexibility more needed than in the Government structures aiming at supporting innovation.

Rigidities are created by a number of factors, which affect institu-tions in different ways. These factors include:

-- Bureaucratic restrictions: for instance, the imposition of restrictions on recruitment across the entire public sector has resulted in some unfortunate distortions in the development of key technological support institutions in Ireland. These institutions require a more flexible mandate from government to enable them to respond positively to the changing demands placed on them by industry as consequence of rapid changes in technology. In this situation, greater freedom to recruit specialist contract staff is essential.

-- Excessive power: conversely, when State institutions enjoy too much autonomy there is no pressure on their management and staff to adapt their attitudes and practices to fulfil better the requirements created by technological change.

-- Lack of information: the decision-making process as well as the arbitration process needed at the government and parliamentary level is hampered by lack of information. Relevant industrial statistics are often not available to decision makers at government level. In this regard we cite below in this Chapter information on the costs of job creation which should, we feel, be better known by those in authority in the area of financial allocations. Moreover, we feel that reports prepared on economic and technological issues should be widely disseminated for public discussion. In this way, agreement on adjustment and adaptation of existing structures will be more easily achieved.

Our conclusion therefore would be that the flexibility needed to fulfil the demands on supporting institutions in a dynamic society cannot be obtained without major changes in the guidelines for these institutions and in their regulations. It is recommended that these guidelines should all aim at decentralising responsibility within Irish administration and conferring a high degree of freedom on public institutions to choose ways and means to solve problems. But at the same time the government should fix clear guidelines and criteria against which the work done by these organisations would be judged.

The next section is devoted to the central concern which should inspire government policy.

Support investment in "brains" rather than in fixed assets

Modern development relies more and more on investment in "brains". Investment in skills, information and knowledge (scientific, technical, design marketing, financial, etc.) is crucial to firms in all industries.

The development of international markets is characterised by more customised demand and the fragmentation of formerly mass markets. Price is no longer a dominant factor. This leads to the need to pay much greater attention to qualitative aspects. Factors such as design, technical excellence, performance in matching customer needs, consistency, reliability, etc., are now essential factors which determine purchasing choice.

Moreover the development and diffusion of the new technologies is leading to a considerable intellectualisation of productive activities: the microprocessor is probably the best example of this intellectualisation process.

Advanced firms in the high technology area invest as much in research and development, marketing, training and software as in buildings, machinery and other fixed assets.

In Ireland, some 80-90 per cent of government support to industrial development amounting to some Ir£750 million per year (if tax relief allowances are taken into account) goes to investment in fixed assets and only 10-20 per cent to investment in "brains". This proportion has not changed over the last ten years, despite (recent) commitments to re-allocation in this regard.

For this "brain investment" in industry, focal points for Irish innovation strategy should be in the short and medium term:

-- Applications of new technologies;

-- Development of market intelligence;

-- Upgrading of the workforce;

-- Enrichment of management capabilities; and,

-- In particular quality improvement in products and services.

This is supported by a well documented study recently published by the Irish Management Institute (9) which argues that the quality improvement of product lines leading to increased market share is the most effective means to enhance the profitability of Irish firms. Conversely, it advises that Irish firms should avoid falling into the "investment trap" which can be created in particular by costly "high tech" facilities and production processes.

To sum up, Ireland should -- as a key element in its development strategy -- follow, with determination, the general trend observable in the OECD area towards knowledge intensive industries, but in focusing on those factors which are the most relevant to her needs and capabilities.

Indeed, it would appear more profitable for the State, and for the country as a whole, to put its money into schemes aiming at developing jobs by adequate investment in training, product development, technical assistance, information, market intelligence, rathan into schemes supporting investments in fixed assets. The cost of one job created or sustained in the former type of schemes appears less expensive than the cost of one job created or sustained by direct grants provided to firms for investments in fixed assets. Some indicative examples came to our notice during our review mission including:

-- The Product Development Scheme of the Training Authority (AnCO): 467 persons have benefited from appropriate courses (20 weeks) during the last year, 80 per cent have been placed in industry; they have generated about Ir£10 million in sales; the cost of training for one job created by this means is estimated to be Ir£2 500;

-- The Youth Employment Scheme developed in the year 1984 by the National Board for Science and Technology and the Institute for Industrial Research and Standards through which 80 persons have been placed; the cost for one job created is estimated to be Ir£4 350;

-- The Institute for Industrial Research and Standards, whose Food Division in 1984 provided technical support to about 10 projects from industrial firms: about 100 jobs were created or "saved" by this mean, the cost per job -- as fees paid by the firms to the Institute -- was approximately Ir£1 000; these projects generated Ir£14 million in sales;

-- The Department of Labour Enterprise Allowance Scheme has in its first year enabled 5 000 unemployed people to start their own enterprises. The enterprise allowance, which runs for one year, is per week Ir£30 for single people and Ir£50 per week for married people. One fourth of the start-ups have been manufacturing enterprises;

-- The Irish Export Board (CTT) has invested Ir£50 000 in developing designs and models for a group of enterprises in textiles and clothing (by hiring US designers): about 1 000 jobs have been sustained by this means making the cost of one job Ir£50.

Although impressionistic, these figures are impressive, even if provision is made for State support given to further investments in fixed assets associated with these investments in brains. The figures quoted should be compared to the estimated cost of jobs created or sustained through the usual industrial development grant schemes. In the latter case cost ranges from Ir£15 000 to Ir£30 000 and sometimes more. Some recent industrial projects in the high technology area have attracted even higher State support than the figures quoted above. This is a disturbing development and a trend that must prove economically unsustainable in the longer term.

As it is clearly more profitable to invest in "brains", a deliberate and sustained reorientation of the grant giving system should be implemented. This applies also to the share of education, research and information in the overall State budget.

IV. SUPPORTING ENTREPRENEURSHIP AND INNOVATION

Much has been said about the apparent insufficiency of entrepreneurial spirit in Irish indigenous industry resulting from historical and structural factors. We are hesitant to follow the conclusion occasionally drawn that the role of the Government must therefore be paramount in the development of this segment of the national economy.

The accomplishments of Americans of Irish descent in the USA as well as Irish people elsewhere, the many dynamic qualities displayed by the Irish in all walks of life when given the right context and opportunities make us confident in Irish entrepreneurship when encouraged by more favourable general conditions and changed attitudes.

Improving social climate for entrepreneurship

The research study published by the Industrial Development Authority entitled Enterprise -- The Irish Approach (10) clearly outlines some of the issues involved in improving the social climate for entrepreneurship.

It emphasizes that the current climate for entrepreneurship seems to be an inhibiting factor in entrepreneurial development in both existing and new enterprises. There is a low credibility in Irish culture for the entrepreneur. While things are beginning to change this is a slow process. To accelerate the process, the report shows the need to promote the image of the entrepreneur throughout society.

The study depicts a number of key features of the Irish entrepreneurial personality, which one found also in other countries, challenging most of the popular assumptions about entrepreneurs:

-- "Entrepreneurs are not solely motivated by monetary considerations; their major concern is often with the provision of a good product, service or process. Money is a very important factor, but it is used as a tool to achieve aims and as a means of keeping score;

-- Entrepreneurs feel a personal obligation to their customers and employees as well as a social responsibility to national development;

-- A critical factor for success is the support they receive from those who are very near to them such as friends, peers, and more particularly spouses; there is little evidence of broken homes or disrupted families;

-- The enterprise is a collective being rather than an individual undertaking. Most successful firms are those which develop as partnerships between complementary personalities bringing together their skills and sharing responsibilities;

-- Entrepreneurs consider the development of their enterprise as a life-long learning process. They learn by failure, and suspicions directed at those who have failed therefore act as inhibitors to enterprise formation and growth."

These features make a portrait of the entrepreneur which should be highly acceptable in Irish culture. Of particular importance is the promotion of the "cult" of entrepreneurship in basic education, including the teaching of practical business economics, with case histories of successful Irish businesses and business personalities. Moreover, the educational system should put less emphasis on security and the need for a safe job. Business formation as a career path should be given more consideration. A number of educational authorities are beginning to institute such an approach, but much clearly still needs to be done.

Moreover, given the general level of maturity of the larger Irish firms there is an obvious need to change the corporate climate in a direction of encouraging and facilitating intrapreneurship, i.e. entrepreneurial activity within the company. The "Company Development Plan" approach recently introduced for government aid to industry should be adapted to develop this.

Direct supports for innovators

For some years, Irish institutions at both the central and local level have developed many initiatives to support innovators. These have borne fruit as was made clear by visits we had to a few small and medium-sized enterprises (SMEs) during our stay in Ireland.

Although these had primarily their business in traditional trades such as food, metal (foundry and sub-supply of spares) and clothing, they demonstrated an ability to earn a profit and to increase export. They all had benefited from governments grants to establish their buildings, and there is no doubt that reasonable rent for appropriate buildings is an important factor in the progress of these firms, e.g. an abattoir in Limerick, which is improving its bacon export business rapidly after moving from an outdated location.

The fighting spirit, low overheads, consciousness of quality and precision in delivery which characterised these entrepreneurs might serve as an inspiration to other Irish employers who apparently have not fully realised what these factors mean for success or failure. Further, especially from a clothing firm visited, it was clearly demonstrated that industrial design combined with proper management and training of the staff, even in this difficult area, may lead to success and increased employment.

These examples underline the fact that there is still much progress to be made inside traditional trade, which to a great extent is based upon Irish raw materials and traditional skills. Therefore there must be a continuation of the rather comprehensive programmes of assistance for small scale start-ups including industrial premises, estates, advisory services, grants for product development and training, linkage service to foreign owned companies and industrial liaison activities which are of key importance for raising awareness for innovation in SMEs.

But in general, we advise that less support be given to investment in fixed assets and more attention to improving the staff skills, to the ability to implant new technology and convert it into new products and processes and to handling export problems.

The comprehensive service given in the Mid-West region under the Shannon Development Company umbrella is of great value for talented entrepreneurs who have a bright idea, but often a lack of knowledge concerning marketing, financing and general management. As it is practically impossible for new starters and small firms to know and profit from the great variety of support they may obtain from the many agencies and institutions, we believe that the Shannon experience should be studied with the aim of decentralising and co-ordinating locally the different types of support.

We welcome the recent establishment of the network of "one stop shops" where information on all industrial supports is available under one roof. We also welcome the devolution to the regions by the Industrial Development Authority of decision making powers on small industry grant applications. These measures are, however, only stepping stones towards the greater decentral- isation which is desirable. Consideration should be given to regionalising with appropriate co-ordination of all innovation-related services and grant schemes addressed to small industry including technology advisory services of the IIRS, R&D grants of the IDA, marketing support programmes of CTT, and services of the Industrial Training Authority and National Manpower Service.

A degree of regionalisation has already occurred in these services but in several cases the number of people based in the local office is totally insufficient to cover its catchment area. The decisions should be taken where they have to be implemented and as many locally based experts as possible should be involved in the decision making boards concerned. The composition of the present County Development Teams may offer a suitable model in this respect.

Integrated support to inventors and other entrepreneurs should be continued, e.g. through the Innovation Centre which in spite of a relatively short working period has obtained promising results compared with similar institutions overseas. The Entrepreneur Intern Scheme and the courses for new entrepreneurs are judged to be of high value for those attending these programmes.

Finally we feel there is a need to expand schemes permitting entrepreneurs to explore opportunities abroad. Since the entrepreneurial spirit is the product of both economic and intellectual stimulation and bearing in mind the limited size of Ireland as a market, it is clear that Irish indigenous industry should constantly be encouraged to explore opportunities abroad. Much is already being done in this respect, notably by the Irish Export Board. Organised study trips abroad by groups of industrialists, particularly to highly competitive markets with high purchasing power, can be rewarding in terms of found opportunities and development of management capability in general, as a result of contact with foreign industrial counterparts. Given the close ties with America, such an approach should in fact offer to Irish industry greater chances for insight into the workings of a dynamic economy than might be the case for others.

Equity investment and sense of responsibility

Government efforts to enhance the financial environment for entrepreneurs and innovators have almost exclusively been based on the provision of grants. This is inadequate and even counterproductive for promoting an entrepreneurial spirit in both new and existing firms.

The IDA's fixed assets grant policy should be re-examined in favour of schemes more apt to elicit a greater sense of management responsibility in the individual or corporate entrepreneur: for example, long term loans conditional upon equity investment by existing or new shareholders, equity investment reinsurance scheme, export credit insurance, etc. In other words, there is a need to transfer the role of the IDA in this respect to the private sector, and equity investment by the private sector should gradually take the place of State grants.

Any measure which goes in this direction would be welcome. This is the case of proposals recently made by a National Commission reviewing the taxation system which recommends strong incentives to stimulate private savings in view of future business creation. Another useful measure would be to favour, by appropriate tax rebate, partnership in business creation.

Private equity or other investments naturally seek a return commensurate with risk. Industrial risk expects greater profit than a safe investment. The partial substitution of private capital for government aid grants would have the advantage of better scrutiny of the profit potential of projects and a greater sense of responsibility in achieving such profits. With greater selectivity, greater quality of growth would be achieved.

This is the most important stimulant as the average return on investment rate in Irish indigenous industry is rather low at the present time -- despite a very favourable taxation of manufacturing profits (10 per cent). It is about two thirds of that on government bonds and a fraction of the estimated return of US investments in Ireland. This low profit performance would deserve a study in itself, but for the future it is clear that high profit projects should be given particular attention. Upgrading of quality, concentration on specialities or particular market niches, etc., go a long way in that direction. Moreover, the profit attractiveness of new projects for risk investors must come from potential long term capital gains, requiring a tax treatment of capital gains which recognises the risks involved in investment. This will be discussed below.

Venture finance

Irish entrepreneurs meet difficulties with the traditional banking system, generally considered as risk averse, bureaucratic, badly equipped to assess projects and searching for "physical" guarantees when making loans. This is not peculiar to Ireland and is to be found in most OECD countries, with the possible exception of the United States. In the long run, bankers -- even bound by strict rules of security regarding the money deposited in their hands by clients -- would necessarily be led to consider it less profitable to take guarantees on "physical" assets than to invest in ideas and talents. Meanwhile, in a number of foreign countries, the decentralisation of the banking system, with strong involvement of managers of local agencies, has proved to be an efficient tool to direct support towards entrepreneurs and innovators, providing better assessment of projects and people, and creating a sense of local community development.

The development of venture capital constitutes an important tool for supporting innovative ventures, complementing usual bank support and bridging the gap with the banking system. Venture capital is long term equity investment comprising two inseparable elements: patient money and hands-on management expertise. It complements, but does not substitute for, the entrepreneur. Its rewards come from long term capital gains rather than current income. This requires good project profit performance. Industrial development proceeds strongly where good ideas combine with venture capital in the expectation of high financial rewards and other satisfactions. Not only in the USA, but in other countries as well, numerous examples of such success stories can be found.

At the present time this helpful avenue appears underestimated in the array of measures being applied in the stimulation of Irish indigenous industry, in spite of the considerable progress already made (i.e. Ir£25 000 annual deduction from personal income for private savings invested in stocks). We believe that a further reduction of the long term capital gain tax to no more than 20-25 per cent or preferably a total exemption would be particularly important so as to make venture capital at least as attractive as investment in homes and flats, industrial buildings, and above all, government financial instruments. This may be important not only for attracting private saving but also "institutional" funds (such as pension funds, insurance companies, etc.) as well as large capital-rich companies.

More generally, the risk of a somewhat excessive liberalisation in the taxation schemes is to be preferred to an excessively restrictive application of incentives, unless the objective of actual and substantial industrial development is not to be met. A second point in this context would be the assurance of adequate longevity for the key incentives, such as the Export Sales Relief, Manufacturing Profit Tax, etc.

Venture funds should also be permitted a reasonable operating margin to ensure their hiring of competent management without which the very concept of venture capital loses its effectiveness. Stock options for investee company management as well as venture fund management should equally receive favourable tax treatment.

Capital gains can only be a strong incentive where efficient mechanisms exist for easy marketing of shares in venture-financed companies. If the Dublin Stock Exchange lacks the necessary features in this respect -- although operating as a branch of the London Stock Exchange -- appropriate steps should be taken to remove whatever obstacles or insufficiencies may exist in this respect.

The encouragement of venture capital companies and venture managers serves a worthwhile purpose in constituting a new breed of industrialists and managers, innovation-minded, internationally oriented, accustomed to, and willing to take and manage, above average risks: in a nutshell, a driving force in a modern democratic competitive society. It would seem in the national interest to maximise such incentives, financial and not least social, as would lead to their rapid multiplication.

In the general context of development capital, some recent proposals appear to merit special constructive attention. For one, the National Development Corporation (NDC), as stated in the White Paper on Industrial

Policy, will be largely guided by long term profit objectives, so to speak by the same basic criteria that would guide a private enterprise (such as a venture fund) contemplating a worth-while industrial project. While under present circumstances it is understandable that the initial injections of capital in NDC will be State-originated, the opportunity should not be missed to offer part of NDC's equity to the general public and/or venture capital funds under especially attractive terms regarding issue price, payment terms or taxation treatment. Provided the commercial considerations recalled above are strictly adhered to, such a course would be a golden opportunity to stimulate participation of private initiative in the development of Irish indigenous industry. Ultimately, the majority of NDC's equity would be held by private capital sources, through repeated public issues of State owned shares, with the State retaining a minority interest and a subsidiary role in management.

This pattern could also be relevant to other projects, such as the proposed Business and Technology Centre at Cork with initial funding being provided by a combination of EEC allocations and private local capital, under private industry management of high calibre. While the initiative is still at an early stage, this type of co-operation between partial public start-up money and private capital, providing experienced management assistance for technology transfer, constitutes a formula for the promotion of innovation by private initiative. Private investment in such ventures deserves favourable tax treatment both at the time of the initial investment, as well as on the eventual capital gains on disposal.

*

*   *

As can be seen, most of the measures proposed above for improving entrepreneurial and financial climate are not costly for the government and proceed from appropriate adaptation of regulations.

V. RAISING THE TECHNOLOGICAL LEVEL OF INDUSTRY

In order to raise the technological level of Irish industry, it seems necessary to break out of a vicious circle. On the one hand, industry, particularly indigenous industry, has a low technological capability and does not seem to perceive the need to remedy this; on the other hand, the research system lacks resources and is insufficiently aware of the industrial problems and demands.

It is absolutely essential that means be established to raise technological competence on a nation-wide basis and to relate this to industrial needs. To achieve this, additional resources must be made available. In addition, the present system, which is based on scattered schemes, operating on modest budgets or on a grant or tax relief basis, should be made more coherent and effective.

One should bear in mind that, as exemplified by Japan in the course of its development, increases in R&D expenditure have followed or gone hand in hand with the building up of a technical culture developed through education, on-the-job training, provision of information, technical assistance and a general strategy of transferring technology from overseas.

Promoting technical research and culture in industry

Efforts made in the past by the Irish agricultural sector for raising its technological level might provide a model for action in this area.

The promoting of technology was conceived in a systemic way integrating several components: training, technical services (Advisory Service/ACOT), development work (at the Agricultural Research Institute), basic and applied research (mostly at University College, Cork). Farmers contributed to the financing of research through a voluntary levy (based on milk production) and there was previously financial support from the local authorities for the Extension/Advisory service (this ceased in 1969). This integrated approach, as developed mainly in the 1950s and 1960s, has permitted the Irish agricultural sector (and particularly the dairy subsector) to improve considerably its production and to be internationally competitive, at least within the EEC Common Agricultural Policy.

This approach could be extended to the development of industrial research and culture on a broader scale. The point is to establish a general mechanism which will support massively the development of a nation-wide network of technological services related to the needs of industry.

A mechanism like the levy grant scheme which finances training activities through the Training Authority (AnCO) could be established (but not on the same basis as the AnCO's levy grant, which benefits only one agency, thus reducing the necessary competition between providers of services). Applied to each industrial firm a levy amounting to one per cent of the added value would raise Ir£40 million; this amount should be compared to the current level of R&D expenditures by industry: Ir£36 million in 1983, in other words a doubling of industrial R&D expenditures. An alternative mechanism would consist in special funds on a model provided by the Swedish Renewal funds: 10 per cent of enterprises' profit are compulsorily reserved for a period of three years and tax deductible for investment in R&D and training. In the case of Ireland, this would generate also approximately Ir£40 million.

These sums would not be used to finance activities which are already covered by existing schemes, i.e. mainly: i) internal R&D subsidised by grant assistance or subject to tax relief; ii) training financed through AnCO.

These sums should be utilised for:

-- External R&D projects contracted by firms to outside organisations (universities, research institutes, other firms, etc.);

-- Quality control, design, testing activities, market research, etc., also contracted to outside organisations;

-- Periods spend abroad, e.g. in foreign companies or universities to learn management, etc.; and

-- Expanding present schemes for the placement of technically qualified staff (with scientific and engineering background) in industry on a temporary basis.

A list of designated organisations to which firms can pass such contracts would be established. In some cases, enterprises could group together for supporting projects of significant size and of collective interest.

Such a mechanism could be extremely efficient for developing demand by industry for further technological effort, if the services are well provided. As an example of this multiplier effect, in operating the first year of the placement scheme for young scientists and technologists in about one hundred firms, the IIRS has generated a demand for technical assistance, product development and so on amounting to four times the amount of money spent in running the scheme.

In the long term, if such a general mechanism proves to be efficient, it would be possible to eliminate most of the numerous grant schemes presently operating for such a purpose or to reserve them to start up entrepreneurs (who do not make profit in early years).

The approach adopted in the agriculture sector to the funding of the Agricultural Research Institute seems an appropriate model for other research organisations. The balance between guaranteed resources coming from government support (Ir£25 million) and discretionary resources coming from levies and contracts, including EEC (Ir£4 million), is correct. This division of income sources, "sure" resources representing 70-80 per cent and "variable" resources representing 20-30 per cent, is to be found in efficient research organisations throughout the world. Such a "rule of thumb" would contribute towards establishing a certain regulation in the budgetary allocations for research by the government. It should facilitate the growth of those research organisations which are most efficient, either in serving local industry or in finding contracts abroad in advanced fields, while providing to these bodies the minimal infrastructure in personnel and equipment needed to satisfy properly the demands put on them.

University-industry relations

In spite of its limited size, Ireland has universities with scientists highly respected internationally. This is underlined by the fact that university staff is carrying out R&D contracts for big overseas companies and for the EEC, as well as for multinational companies located in Ireland. As an example, the Trinity College of Dublin in 1983-84 had a research income of Ir£2.4 million, from which over one third was funded by overseas companies and the EEC.

There are also striking examples of teams nurtured within faculties who have gradually established themselves as research enterprises with top world expertise. An example of this is a team of Trinity College, composed of experts in remote sensing by satellites for geological studies, selling their competence to foreign multinationals as well as to developing countries. Similar examples might be illustrated from other third (or "tertiary") level institutions.

Further, Irish graduates in science and engineering demonstrate their qualifications at an international level by the relatively well paid jobs they may obtain when they emigrate, or when working in multinationals located in Ireland.

Indigenous Irish industry is using this eminent resource of highly skilled manpower only to a very limited degree. As with the relatively low technology level, this is often related to the small size of companies and the lack of R&D, and traditional attitudes in Irish indigenous industry. That situation has also to do with the rather low output of graduates in relevant areas such as engineering and science compared with other industrialised countries, -- and perhaps with the traditional "ivory tower" image of universities.

Further development of co-operative links between the higher education sector and industry must be seen as a priority policy objective. Many initiatives have been established in recent years to narrow the gap. The National Board for Science and Technology (NBST) has for a number of years been supporting this co-operation through a variety of funding and linkage mechanisms. These initiatives are demonstrating the value to industry of using highly skilled manpower, and also demonstrate changing attitudes in the universities, from seeing themselves exclusively as members of a scientific community whose over- riding, common aim is the advancement of knowledge, to seeing themselves as a vital base for Irish industrial development.

The problems related to this theme are described in many reports, e.g. in Strategy for Industrial Innovation: The Role of Third Level Institutions (11). Based upon experiences in Ireland as well as in other industrialised countries, it is obvious that the stimulation of university-industry relations must be built on mutual interests of both parties.

In addition to financial measures discussed above to increase money spent on R&D projects of importance to Irish industry and to stimulate co-operation with universities, a number of institutional and regulatory improvements are also needed. As illustrative examples, not as a complete list, are recommended:

-- Further development of efficient liaison between universities and industry, with a view to identifying education needs in consultation with industry, and to promoting an awareness among employers of the available expertise; this can be done by the establishment of advisory bodies to universities/faculties including industrialists, which are not aimed at governing research and education but with a role in evaluation of the activities and in discussion of aims and means, according to the future needs of industry and society in general;

-- The establishment of a Teaching Company Scheme, to market the third level institutions' expertise and facilities to industry, to establish joint ventures, and to encourage students, teachers and industry to have students work as trainees in industry and to solve problems of interest for industry in their projects; to stimulate third level institutions to run short courses and other arrangements with the aim of upgrading the knowledge about new technology in Irish industry;

-- Developing flexible guidelines for university staff to perform contract work in co-operation with industry. Such guidelines might have a motivating element for researchers as well as for institutions. If such common guidelines are not established, many different practices will show up, including "under-the-counter" contracts, which are destroying proper relations between industry and university. The so-called "Devlin rules", therefore, have to be radically revised. The regulations bearing on the activities of the Regional Technical Colleges which seem to hamper initiatives in working with industry should also be reviewed; we understand that these are currently under review;

-- Encouraging the establishment of more facilities to improve university-industry co-operation. Industrial science and technology parks as in Limerick, or "incubator factories" on a smaller scale as in Galway, alongside academic institutions are mechanisms which may be favourable. Another would be the creation of units with a self-governing status attached to universities and being responsible for contract research in certain areas or at a multidisciplinary level. Inspiration might be obtained from the Norwegian institution SINTEF which is a technology commercialising body attached to the University of Trondheim;

45

-- Supporting arrangements where researchers and industrialists together visit advanced overseas industries and institutions to study specific fields of interest for Irish industry, and to facilitate co-operation between Irish universities/industry and similar institutions/companies in other countries to gain profit from the EEC research programmes (ESPRIT, BRITE, EUREKA, etc.);

-- The encouragement of appropriate concentration of expertise in particular colleges or associated with particular colleges so as to ensure cost-effective and viable levels of efforts; and the encouragement of greater co-operation between colleges in the selection and use of these concentrations of expertise.

Foreign owned industries in Ireland

As a vital point of its industrial strategy Ireland, since the 1950s, has attracted foreign investment through a broad and generous range of fiscal and financial incentives. Until the end of 1980, profits derived from manufacturing exports enjoyed complete tax exemption, and some companies may continue with this until 1990. Incentives offered include a tax as low as 10 per cent on manufacturing profits, depreciation allowances of 100 per cent against tax, capital grants up to 45 and 60 per cent (mainly in the West) for buildings, machinery and equipment, and labour training grants up to 100 per cent as well as grants of up to 50 per cent of the cost of R&D.

As a result of this strategy there has been a rapid growth in the number of foreign based companies in Ireland, accounting for about 80 000 jobs, i.e. 37 per cent of employment in manufacturing industry. The main part of the production from these firms is exported and they are responsible for about 65 per cent of total exports of manufactured products in Ireland. It should be added that Irish based subsidiaries of multinationals give very high rates of return, i.e. about 25-30 per cent according to available statistics.

The strategy of government has succeeded insofar a process of structural change has started in industry in Ireland by attracting advanced, internationally research-intensive high technology enterprises. In the Western region this has been of decisive importance for creating a new era based upon new technology. Industrial production has grown faster than in any other European country, mainly due to foreign owned industry, especially in the electronics and chemical sectors.

But from other view points this heavily supported development has some problems, e.g. the often mentioned lack of cross-breeding between foreign and indigenous industry. In other words, Irish industry has developed a duality. The foreign owned industrial sector has expanded rapidly by means of high productivity, modern technology and international exports. Meanwhile much of the indigenous sector suffers from low productivity and low speed in product renewal, implementation of new technology and international marketing.

Ireland needs some high technology input from leading industrial countries, and the very limited home market is not in itself of such an interest to foreign companies that they prefer to place their investments in Ireland without receiving some incentives. In the long run, however, Irish society must seek a better return for its generous terms towards foreign industry establishing in the country.

Logically, the incentives in future ought to be directed more to "brain" development than to fixed assets, as mentioned earlier in this report.

A series of commitments could for instance be imposed as a precondition for any new grant aid to foreign industry. Any firm in receipt of grant aid in future would be obliged to commit itself to, or show a record of, doing some of the following things:

-- Establishing an R&D capacity in Ireland (to ensure that the subsidiary is not exclusively engaged in routine work based on imported components and R&D work executed abroad);

-- Purchasing, within three years of start-up in the country, a defined proportion of its raw materials, parts and services from Irish sources;

-- Providing on a contract basis assistance to indigenous firms in the areas of quality control, productivity, product development and general management;

-- Using a defined proportion of their profits for contract research in Ireland;

-- Hiring young Irish technicians and/or engineers and organising a training/personnel development programme for all employees;

-- Sharing the expertise of key management personnel (be they Irish or foreign) with students in second and third level education establishments by acting as occasional instructors or guest lecturers.

Conditions like these may be seen as a burden for a foreign investor, if he just wants to profit from the Irish grants in a short period. But if the investor has serious ideas of a long term investment, he certainly might be aware of the mutual interest in improving the indigenous industry as well as the labour force.

In the Shannon and Limerick area, some of the most progressive multi-nationals have already voluntarily taken up some of the activities mentioned above. And it is open to question if foreign investors who are not ready to accept such conditions are appropriate objects for the comprehensive investment that Irish society puts into foreign companies when they locate in Ireland.

VI. INTENSIFYING THE EDUCATION EFFORT

The formal education system

In traditional indigenous industry a rather low proportion of the staff are at the technician, professional and managerial level, while the big majority are unskilled/skilled workers and clerical staff. In modern high technology industries the estimate is that over 50 per cent might be technicians, or higher grade, and less than 50 per cent production workers. If such a 50/50 base is used and Irish industry is going to compete by means of technology, quality and good management, employment in Irish industry in the year 2000 of 320 000 will include at least 150 000 persons with a professional background above skilled workers' level. Among these there will be a minimum demand of 50 000 with a third level background in engineering, science, economics, marketing, etc. Contrary to this, demand for semi-skilled and unskilled workers will be lower than now, because many routine jobs are expected to be superfluous.

Various studies indicate that, proportionally, Ireland has only one third of the number of engineers and technicians in industrialised countries such as Japan, France, the USA and Sweden. This might be a result, as well as a cause, of the generally less developed indigenous industry.

The supply of engineering students has increased by about 10 per cent per annum over the last three years. The engineering graduates' output in the period 1981-90, as estimated by the Higher Education Authority in 1982, is shown in Table 7.

It will be apparent from the above remarks and statistics that there might be a severe mismatch between the output and the potential demand, if the production of engineers and other highly qualified employees is not rapidly accelerated, especially in areas such as: electronics, data processing, production engineering, materials science, chemistry, biotechnology, product development, strategic planning, economics and marketing.

In the "Programme for Action in Education 1984-87" many positive initiatives are described. It is essential that they are brought to reality, and if possible accelerated. The fact that annual capital spending on education is planned to rise by Irf32 million or 38 per cent in the period 1984-87 is very constructive. The establishment of the National Institutes for Higher Education (NIHE) has demonstrated that ideas and plans may be realised with a high degree of success.

Table 7

ENGINEERING GRADUATE OUTPUT -- 1981 (ACTUAL) AND 1982-90 (ESTIMATED)

Type of course	1981	1982	1983	1984	1985	1986	1987	1988	1989	1990
Chemicals	59	38	49	64	73	73	73	86	86	86
Civil Engineering	173	189	177	219	207	215	215	215	215	215
Electrical	213	225	296	346	334	347	357	409	409	409
Mechanical	193	143	226	268	288	292	380	405	410	410
Agriculture	12	10	15	7	16	16	16	16	16	16
Other	6	4	42	56	51	81	116	194	194	194
Total	656	609	805	960	969	1024	1157	1325	1330	1330

We strongly recommend that there should be an increase in the intake of engineering and technology students by 25 per cent per annum over the next 5-10 years, and at the same time other university studies should be geared to the future needs of a society based upon advanced technology and international trade.

In developing the third level of education, inspiration could be obtained from NIHEs' curricula which have proved to be efficient and also from the experience of the "Bolton Street College of Technology", the diplomas of which seem to be very highly regarded.

Parallel to this there is a need for expanding the quantity of technicians and skilled workers educated in the Regional Technical Colleges and for continuous upgrading of the courses offered in this regard. There is an enormous challenge also in creating training possibilities for grown-up people to master the new technology and other future qualification demands.

It is recommended that industry in co-operation with universities and other relevant institutions make a thorough survey of future qualification demands, to meet their needs, in order to give a base for appropriate adjustment of the educational system. This might lead to changes in curricula with the consequence that college lecturers must be given the opportunity to improve their knowledge and teaching methods.

Science and engineering studies, but also studies in humanities, sociology, psychology, teaching, economics, business management, etc., should be geared to future jobs in the private sector. This sector has a need for well trained staff with international -- and not exclusively an Irish (or British)-horizon. Although science and technology are vital ingredients in the innovation process, it is essential, too, that "soft sciences" are included to ensure well balanced technological development and to improve working relations with foreign countries, which are going to be the primary customers, competitors and partners.

Consideration should also be given to how the use of TV, video and computer assisted learning could be instrumental in this immense task of upgrading knowledge in the workforce.

A thorough examination of education at primary and secondary level is at present taking place conducted by the Irish Curriculum Development Board. Many sources indicate that this part of the educational system is far from optimum in curricula as well as in teaching methods. The fact that women are not generally educated in science and technology needs to be changed. Furthermore some practical experience in technology should be of great value to students in all secondary schools; familiarity with computers by a "hands-on" approach is of great importance for their future studies and working life. The Danish education authorities have developed some interesting models in this respect.

The idea behind such a massive action to improve the qualifications of the "young Europeans" as well as the existing labour force is that the only real competition factor for Ireland in future will be well trained, highly efficient, self-reliant manpower. The great proportion of young people in the population should be regarded as an opportunity to build an innovative nation.

If education is not broadly upgraded and if a more innovative climate is not created, there is a severe risk that a generation of young Irish men and women will be frustrated by never having adequate education and jobs. People who are not given opportunities for adult education and retraining will run a greater risk of unemployment. This certainly must cause social and political problems of a magnitude which is totally unacceptable.

The role of the social partners

In this broad upgrading and diffusion of technology throughout Irish society, Industry Federations and Trade Unions have an important role to play.

Based upon experiences from other countries it is recommended that major attention be paid to the role of trade and industry associations such as the Confederation of Irish Industry (CII) and its affiliates as they have a natural interest in assisting their members to improve business results. The organisations often are trusted much more by their members than are government offices. These organisations know the essential problems and know the language of the enterprises. But it may be possible to motivate CII and similar organisations to strengthen their efforts for innovation alongside their general activities to improve the business climate in Ireland. For instance they may well be able to run services for the benefit of their members at a reasonable cost.

Employers associations and trade unions also have an important role in the wider development of society. It is recommended that the two social partners should establish a clear framework within which the technological improvement of Irish industry might be supported to their mutual advantage. The existing tripartite mechanisms provide some opportunity for this. Experiences from other countries, e.g. the Scandinavian countries, might provide an inspiration for a more developed approach for additional efforts by the two social partners themselves. In these countries social partners have obtained really good results from co-operation, e.g. in preparing and running courses, preparing reports and arranging conferences on new technology and its consequences, establishing consultant services concerning productivity, etc.

It is evident that many employees with a weak professional background feel unsafe in a turbulent society, which demands flexibility and a range of new qualifications. If a mutual understanding exists between employers and employees concerning the perspectives and problems raised by technological change, its implementation is rather easy, especially when the plans are discussed before execution and adequate training offered to the staff in due time.

Following this line of thinking some consideration might be given to how the climate in the labour market might be improved so as to favour the introduction of new technology, increased productivity and quality, which are all matters of mutual interest for employers and employees.

VII. SUSTAINING THE VISION

As said in the introduction, central to the innovation policy is a vision broadly shared by the population. No country can progress without a vision of its future.

In an agrarian society -- which Ireland was up to the fifties -- the source of wealth and power was the land. In an industrial society the source of wealth and power was capital, and Ireland has successfully attracted it to boost its development during the last thirty years. Today new technologies, and particularly information technologies, lead to a massive investment in knowledge, and the new source of wealth and power is the minds and brains of the people; in other words, its human resources, properly developed. Along those lines, Japan, for instance, is building up a nation-wide vision based on the concept of 'the communication society'. This is a real driving force in Japan's present success. Pioneered by its Research and Industry Ministries, France has developed an even more potent vision: The Creative Society (as opposed to the Production Society) (12).

Ireland can sustain a vision for its future on a related concept -- as an example to developing countries and regions around the world.

In other words Ireland could become a "model for development". This model would be based on the following factors:

-- Massive and continuous development of its human resources;

-- The best use of new technologies for fertilizing local resources with ambitious goals for employment creation;

-- Creation of an entrepreneurial climate using cultural features geared to private and local initiatives and plug-in international networks of knowledge and technology;

-- The search for excellence and sense of quality in all areas of national life ranging from industrial production to public management.

This model would certainly be attractive for many of the less developed regions in Europe, the Pacific area -- the new centre of gravity of the world economy -- where there is an important Irish connection, and the Third World, with which Ireland has established a number of links through humanitarian actions. Such a global project would generate widescale discussion among the population. It would help in opening minds and planting seeds for a powerful drive forward in economic development.

But a vision does not develop itself in a vacuum. A vision cannot be created by specific institutions, even those entrusted to look towards future. Developing a vision is a gradual process which involves society as a whole. It needs credibility and practice -- people see when they practise seeing. So things should be made visible. Some remarks can be formulated:

-- Expose to the people the country's concrete achievements: as a starting point, a broad campaign should be launched in the media, in schools, in local communities, etc., to show illustrative achievements already made by Irish entrepreneurs and communities along the model of development described above;

-- Develop on the media side better coverage of economic and industrial development issues. This would have an encouraging effect on the climate for innovation and entrepreneurship in the country. There is a real need in Ireland for widely disseminated, accurate information on technology, marketing and finance issues. Such information would be of direct use to people in industry in making decisions on questions such as pay claims, investment plans and technology acquisition. More generally, the media itself should attempt to present more forward looking and outwardly oriented coverage, stimulating the economic ambitions of their audience and opening their minds to the positive aspects of new technology;

-- Mobilise people on concrete projects of national significance: for a start a nation-wide programme of quality improvement could be launched to address this issue in a new development era. This would help in motivating people and focusing technological effort on visible achievements directly relevant to their own lives. An enormous potential exists in Ireland for quality improvement in every walk of life, and notably environment, urban planning, workplaces, consumer goods and services. This would constitute a powerful tool to upgrade in a relatively short time a large segment of industry, while making it aware of the possibilities for productivity increases and gains in market shares. Employers' confederations and trade unions, as well as education bodies and mass media should be involved. Relevant parts of the administration including in particular those bodies in charge of procurement policies and local authorities should also participate in this exercise. Such a programme should not involve extra expenditures but should use existing resources through appropriate co-ordination and reorientation;

-- Utilise local authority budgets for building up local infrastructures appropriate to this new technological age: "incubator" industrial units, innovation workshops, computerised local libraries with access to relevant databases for enterprise development and creation, etc.;

-- Involve, in a pervasive way, social sciences in all economic life, e.g. for developing new economic models, new national accounts integrating investment in human resources. Ireland has a good research tradition in human and social sciences. Making the best use of those scientists consists in associating them closely with concrete technology projects. This is the best way to introduce technology in society. Here again previous experience in agricultural development with farmers' communities may show the direction; and

-- Above all, provide examples: this is why government should make clear gestures committing itself to the basic principles suggested in Chapter III of this report concerning the role of private initiative, the priority to investment in brains, and the bottom-up approach. That would help in creating a kind of culture shock that the country may need today to release and develop its potential -- which is considerable.

NOTES AND REFERENCES

1. Programme for Economic Expansion, Government Publication Sales Office, Dublin, 1958.

2. Economic Survey -- Ireland, OECD, Paris, April 1985 (pp. 47-49).

3. A Review of Industrial Policy, National Economic and Social Council, Dublin, 1982.

4. White Paper on Industrial Policy, Dublin, 12 July 1984.

5. Building on Reality, 1985-1987 (Three Year Plan), Dublin, October 1984.

6. Report on International Competitiveness, European Management Forum Foundation, Geneva, 1985.

7. Irish Journal for Business and Administrative Research, Dublin, Spring 1985.

8. PETERS and WATERMANN, In Search of Excellence, Harper and Row, New York, 1982.

9. CARROLL C., Building Ireland's Business, Perspectives from PIMS, IMI, Dublin, 1985.

10. Enterprise -- The Irish Approach, published by the Industrial Development ment Authority, Dublin, 1982.

11. Strategy for Industrial Innovation: The Role of Third Level Institutions, published by the Confederation of Irish Industry, Dublin, 1982.

12. Cf. Rapport sur l'Etat de la Technique, Centre de Prospective et d'Evaluation, Paris, 1985.

Annex I

ACCOUNT OF THE REVIEW MEETING

The Committee for Scientific and Technological Policy met in Dublin on 14th October 1985 for the review meeting. The session was chaired by Mrs. Hommes, vice-chairman of the Committee. The Irish Delegation was led by Mr. O'Connor.

List of participants

Chairman	Mrs. R.W. Hommes	The Netherlands
Examiners	Mr. M. Knudsen	Denmark
	Mr. J.-E. Aubert	OECD Secretariat
Irish Delegation	Mr. C. O'Connor	Assistant Secretary, Department of Industry & Commerce
	Mr. M. Manahan	Department of Industry & Commerce
	Mr. S. Aylward	Department of Industry & Commerce
	Mr. T. Higgins	National Board for Science & Technology
	Mr. D. O'Doherty	National Board for Science & Technology
CSTP's Delegates	Mr. G. MacAlpine	Australia
	Mr. G. Kint	Belgium
	Mr. M. Koskenlinna	Finland
	Mr. Giraudet	France
	Mr. G. Houttuin	The Netherlands
	Mr. K. Roed	Norway
	Mr. A. Rimas	United States
	Mr. M. Paillon	Commission of the European Communities
OECD Secretariat	Mr. J.D. Bell	Head, Science & Technology Policy Division

Introduction

Mr. Cormac O'Connor speaking on behalf of the Minister for Industry and Commerce thanked the OECD and the examiners for their report which he had found highly interesting and stimulating. It complements a series of OECD studies which had contributed in the past to the development of science and technology in Ireland, e.g. the science policy review and the scientific and technological information policy review both published in 1974. Mr. O'Connor expressed his pleasure that Ireland had been chosen as the first candidate for this new series of innovation policy reviews.

He stressed that the broad thrust of the recommendations in the report and many of the individual recommendations commended themselves to his Minister Mr. Bruton; they would be referred to publicly when the Minister presented the NBST Innovation Awards, as part of the "National Innovation Day" activities being organised around the examiners' report (on this 14th October 1985).

The Minister was particularly in agreement with the view that Ireland should get away from the mentality that "everything can be solved or promoted by grant assistance" and that "private equity investment should gradually take the place of State grants". In relation to this, Mr. O'Connor felt mention should be made of the Business Expansion Scheme. This should stimulate private equity investment. The Minister would have liked the report to put greater stress on profit sharing with workers, as a way of accelerating the process of innovation in firms and of helping to ensure the active partici- pation of the workforce.

The need to support investment in "brains" rather than in fixed assets and to implement a major shift in emphasis in favour of indigenous industry has been urged by the White Paper on Industrial Policy. These principles also -- clearly set out in the OECD report -- should guide further government support to industrial development.

He noted that the report argued strongly for decentralisation. However the Minister feared that decentralisation of power without decentralisation of financial responsibility might be a dangerous development, and he considered that there was a need to devise an acceptable accompanying system of local taxation if a policy of decentralisation was to be effective.

The Minister endorsed the need for a vision of the future while expressing some reservations about the specific targets proposed for the year 2000:

-- Increasing the number employed in industry and related services by 100 000; and

-- Doubling Ireland's share of the world market for industrial products.

The State agencies under the Minister's aegis would be asked to study the OECD report and indicate how its goals might be achieved within the constraints that the report itself sets out on the "stabilization and possible reduction of overall public expenditures".

The discussion then commenced with the participation of representatives of other Member countries as well as the examiners. Its main conclusions are summarised under three headings:

-- Design of industrial strategies and specific innovation measures;

-- Education and the development and utilisation of human resources; and

-- General environment and cultural context for innovation.

Industrial strategies

It was noted that in proposing a vision for Ireland the report gave precise objectives in terms of exports and employment. But the report did not provide clear indications on the industrial strategies to be pursued to reach these objectives. The examiners were asked to clarify their position on this issue.

The examiners answered that rather than designing detailed industrial strategies it might be more important for the government to establish a framework in which the creativity of enterprises can flourish, and thus to eliminate obstacles and disincentives to innovation. When the "climate" is favourable, the enterprises themselves find the technologies and markets in which to invest. In a country like Denmark the government does not devote much effort to industrial planning. On the contrary, it attempts to create a climate of iniative and optimism and put the emphasis on "infrastructures" such as education.

However, a few general principles should be kept in mind in implementing industrial strategies and in providing government support:

-- Avoid overemphasizing the high technology sectors per se, but build on existing competences, and take advantage of strengths existing in traditional sectors, and catalyse through the diffusion of new technologies;

-- Rather than providing support to broad areas or sectors, identify niches where there are dynamic and competent entrepreneurs with products enjoying a potential competitive edge;

-- Be pragmatic: for instance an easy way to become an exporter is to sub-supply to foreign firms; there are possibilities for other forms of linkage between large and small firms. A number of State agencies could do more to help Irish firms to follow such a path.

A key element in developing industrial strengths and technological competence remains the establishment of centres of excellence and expertise. Industrial success, ultimately, depends on the development and application of talent. There is a need to build such core skills in some key areas; agro-food and related biotechnology, for instance, might receive attention in view of their importance to the Irish economy. These could then be built on in the manner in which Finland has developed from its base in the processing of wood and related products.

Human resources

The report stresses the importance of developing human resources. Questions were raised as to how this might best be undertaken.

In the view of the examiners, Ireland was lucky to have a high birth rate. This creates a difficult challenge in terms of job creation, but people also constitute a considerable resource. Ireland should take advantage of this situation in achieving its educational and economic objectives.

The view was expressed that the primary school may be the weakest part of the education system. All pupils should have a basic understanding of science and technology and its place in economic development. This applies to both boys and girls, the latter having been relatively neglected. Particular attention should be given to developing computer literacy. An entrepreneurial attitude needs to be engendered among young people.

A major effort is also needed to develop the basic skills of the workforce. Most of the workforce in Irish industry is either unskilled or semi-skilled. This situation must be changed if Ireland wants to compete successfully in world markets. It does not take a major effort in time and money to update skills and familiarise workers with new technologies, such as computers. In many cases well designed short demonstration programmes are sufficient. Existing schemes operating through AnCO should be carefully reviewed and improved. Trade unions should also be made more aware of their role. They need to adopt more progressive attitudes, following the examples of countries such as Denmark or the United Kingdom, where trade unions invest in venture capital funds for instance.

The intake of engineering students in Irish universities should be greatly increased as underlined in the report. Some people may fear a brain drain, as these highly educated people may not find jobs in the country. But that puts a question mark over the nature of the education which is provided. Education should be properly oriented towards the needs of indigenous industry and should not be guided by purely academic criteria. Greater priority should be given in the curricula to practical training in industry work, as is done for instance by the National Institutes for Higher Education.

A cultural change

While certain elements of Irish society see the need to base future development on science and technology, to orient industry towards international markets, to invest in education, etc., the absence of an innovation-oriented environment and culture makes it difficult to implement the appropriate measures. Questions were raised as to what could be done to stimulate change at this level.

The examiners emphasized the need to:

-- Encourage rapid change in the general atmosphere through simple policy measures of a financial character, e.g. generous tax relief on private money invested in venture capital operations or business enterprise development;

-- Make achievements visible: promote the innovators, find successful entrepreneurs, dynamic craftsmen or innovative trade unionists who can motivate others, creating a climate of optimism; in the longer term successful innovators should take their place among the heroes of Irish society;

-- Involve politicians through appropriate links and networking with innovative groups. As a means to creating employment and to rejuvenating industries or regions, innovation should be understandable to politicians and attractive to voters; and

-- Introduce flexibility and decentralisation within State agencies. To become more effective a number of these agencies need greater power and responsibility rather than more money. This would help to stimulate an innovative approach within the Administration itself.

It was felt also that the best use should be made of the 40 million people of Irish extraction who are living throughout the world and notably in the United States. A country like Israel has shown all the benefits to be derived from a good mobilisation of its "diaspora". Rather than attracting foreign multinationals to set up factories in Ireland (at an increasingly high cost), it might well be more profitable to establish appropriate programmes and incentives to attract bright entrepreneurs, or scientists, of Irish origin and to create a climate in which they can be as successful as they are abroad.

However the examiners asked whether Irish society really wants to change. The emphasis must be on entering a conscious process of change by both collective and individual actions.

Annex II

BACKGROUND NOTE PREPARED BY IRISH AUTHORITIES

Context

Ireland has been going through the process of reassessing its national industrial development in recent years. Despite high and rising levels of production, productivity and exports Ireland is facing actual losses of employment in many sectors, a rapidly growing labour force and a fight for survival in many of the indigenous firms which still make up 60 per cent of industry.

Policy response

The policy response has operated at two different levels.

At the national level, an overall policy review was initiated by the Government through the National Economic and Social Council -- a forum for the "social partners" much like that in most other OECD Member countries. The Council commissioned an international consultancy firm -- the Telesis Group -- to carry out a fundamental review of national industrial strategy. Using corporate strategic management tools, which, for a small country at least, offer many useful insights into the operation of the industrial system, the Telesis Group made a number of criticisms of the industrial development policies then in operation and some major recommendations on new policy approaches.

This was followed by responses from the Council itself and from the Government in the form of a White Paper on Industrial Policy in July 1984.

All of these overall industrial policy reports recommended a major increase in emphasis on the development of indigenous industry and from the fixed capital to the human capital elements of company operations generally. In other words it is the technological and marketing resources and capabilities of the firm which should be the future focus of financial and technical assistance from the government industrial development infrastructure. The White Paper referred directly to the need to improve product development performance in Irish industry as well as doubling industrial output through the application of new process technology over the 10 year timescale taken. While it did not use the term "innovation policy" as such, the White Paper implicitly elevated innovation into one of the major strands of industrial development strategy.

On the 14th October 1985, the Minister for Industry, Trade, Commerce and Tourism released the first ever OECD Examiners' report on national innovation policy. This initiative had arisen directly from the conclusion of the Dubrovnik Workshop on Technological Innovation Policy in less industrialised OECD Member countries (September 1983).

Concurrently with the process of reassessment at the broad policy level a number of new initiatives and approaches have been tested at a more microeconomic level. During the 1960s and the 1970s the Irish approach to industrial development was characterised by a high degree of pragmatism within a clearly defined strategy of foreign investment attraction and the encouragement of small domestic start-ups. Ireland now has brought this flexibility into its new industrial development strategy. The effect has been to maintain a policy where major efforts are made to use existing resources both more effectively and more efficiently.

Institutional background

A description of the key Government organisations involved in industrial innovation is given in Appendix I. A financial statement of Irish Science and Technology allocations for 1985 is given in Appendix II.

Appendix I

IRISH SCIENCE AND TECHNOLOGY ORGANISATIONS

Institute for Industrial Research and Standards (IIRS)

The Institute employs about 600 people and had a turnover of the order of Ir£14 million in 1985. Its main functions are to encourage and assess the use of science and technology in industry. It is the largest technical services institute for industry and its main activities are the provision of consultancy services, technical information and advice, testing and technical services, and carrying out some research for industry, government departments and agencies.

Much of the Institute's work is undertaken directly for individual clients on a contract basis. The Institute has in the order of 3 000 fee-paying clients every year, representing 10 000 jobs. Fees account for about 40 per cent of total expenditure.

The main areas of activity which support manufacturing industry are: biotechnology including food processing, biochemicals, electronics, engineering, information technology, materials, standards, textiles and timber.

The IIRS is also responsible for drawing up Irish standards specifications. In 1985, a new Standards Authority with an independent board was established, based on the existing IIRS standards activity. The Authority proposes to produce new standard specifications while sales of existing standards are expected to increase.

An Foras Talúntais (AFT) -- The Agricultural Research Institute

AFT had a staff of 1237 and a turnover of Ir£23 million in 1985. Income is received from the Department of Agriculture (Ir£16 million), and from fees and donations (mainly agreed payment of levies in return for the carrying out of research into projects for particular agricultural sectors), investment income, sales of produce and surplus on livestock operations. Its general functions are to review, promote and undertake agricultural research.

Research carried out in the food area is the main area in which the Institute supports manufacturing industry and it accounts for 10 per cent of AFT's expenditure.

Increasing the added value component of the food industry, particularly meat, dairy products and cereals is one of the main objectives of research in this area. Projects being undertaken include the use of dairy by-products in processed meats; soft cheese manufacture; semi-aseptic packaging; examination of potential markets for continental-type cheeses; genetic engineering in livestock and food chilling.

In addition to research, AFT analyses a wide range of samples for industry, such as dairy products, meat and meat products. An information and advisory service is given to industry and research staff give on-site advice on processes used in industrial food plants.

Patent Office

The Irish Patent Office was established in 1927. It employs 95 staff and is publicly funded. It experienced a surplus, arising from fees and sales, over costs of almost Ir£1 million.

The principal functions of the office are the granting of patents, the registration of industrial designs and trademarks and the provision of information in relation to patents, designs and trade markets.

In terms of technological support for industry the Office provides a comprehensive information service to firms comprising legal and technical works of interest, and to persons concerned with designs, trademarks and copyright, including current applications and proposals.

In addition Patent Office staff are involved in the examination of applications from firms for patents, designs and trade marks.

The Patent Office also participates in international matters falling within its remit.

Irish Productivity Centre (IPC)

The Irish Productivity Centre is an autonomous organisation jointly controlled by the Federated Union of Employers and the Irish Congress of Trade Unions. It is financed by a grant-in-aid from the Department of Industry and Commerce, by EEC grants, rent income and fees charged for services provided.

The objectives of IPC are to help smaller businesses become stronger and more competitive and to develop more effective human relations within firms, with emphasis on work organisation and worker involvement. IPC employs a total of 49 staff, 37 of whom are employed in areas related to science and technology activities.

The main activities in these areas are in the provision of business consultancy and advisory services and in carrying out studies relating to the objectives of the organisation. Studies include such topics as the implementation of new technologies and investigations into business start-ups and entrepreneurial activities.

National Microelectronics Research Centre (NMRC)

The NMRC which was established in 1982 is a semi-autonomous unit attached to Cork University College. It operates under an advisory board which includes representatives of the electronics industry, the academic community, and the Department of Industry and Commerce.

The Centre employs a total of 43 staff, 14 of whom were permanent in 1985. Operating expenses were Ir£713 000 in 1985 of which Ir£213 000 was funded from research and development contracts, the remainder coming from the Department of Industry and Commerce (Ir£300 000) and the Higher Education Authority (Ir£200 000). In addition, capital in the order of Ir£422 000, was provided by the Department of Industry and Commerce in 1985.

The NMRC was established to provide the infrastructure necessary to assist the development of semiconductor and microelectronics component manufacturing in Ireland; it is intended as a resource centre with specialised plant and expertise in the field of semiconductor design and fabrication technology. Its primary function is to be a national facility providing advanced training and research in semiconductor and integrated circuit design, fabrication and testing; to accomplish this the NMRC will promote this technology and provide access to it by the higher education and industrial sectors.

In addition to R&D contracts being undertaken on a commissioned basis for Irish and foreign based firms, the EEC and other bodies, the Centre also carries out fundamental research without reference to immediate client's requirements; the elements of the programme are determined within the Centre and in co-operation with other academic institutions.

Kilkenny Design Workshop (KDW)

The main function of the Kilkenny Design Workshop is the advancement of good design in industry and in consumer standards. It offers consultancy services to manufacturers and other organisations in the planning and implementation of design strategy, and provides a comprehensive range of practical design services to fulfil the needs of most business organisations in respect of products, information systems and corporate identity. Activity areas include ceramics, precious metals, textiles, furniture, capital and consumer goods, model and prototype making.

KDW employed 133 staff and had a turnover (exclusive of its retail shop) of Ir£1.1 million, of which Ir£875 000 was received from the Department of Industry and Commerce in 1985. The remainder came from the EEC and the sale of design services.

The level of design in industry is very weak and KDW has considerable expertise in this area, in promoting the design of products with good consumer appeal and good aesthetic qualities.

The Innovation Centre

The Innovation Centre is an autonomous unit of the Shannon Free Airport Development Company and was established in 1980 in collaboration with the IDA, the IIRS, the NBST and the National Institute for Higher Education (Limerick). Its main function is to provide comprehensive product development services for small industry, with a strong bias towards more advanced technology and the creation of projects capable of commercial viability in international markets.

Activities extend from product sourcing -- including licensing -- to product and process development, prototype construction, test marketing, pilot production, business planning and personnel development in these areas.

Following a review of the Centre's activities for the 1980-84 period, it was decided that in future the Centre would focus on assisting the creation of Irish owned, innovative, export-based enterprises.

In order to achieve this objective the Centre has an innovation fund from which it provides "seed" money to assist companies. It also has its own laboratories and workshops and provides hands-on advice and assistance. It had a turnover of Ir£600 000 and employed 10 staff in 1985.

The Centre also tries to secure venture capital for its commercialised projects. As is to be expected, only a certain proportion of the projects which the Centre supports prove commercially successful, but those which do succeed in the market place reimburse five times the Centre's investment in their project.

Microelectronics Applications Centre (MAC)

The Microelectronics Applications Centre was established in 1981 and is controlled by a board of directors representative of its founders; these are the IDA, SFADCo, NBST, the National Institute for Higher Education (Limerick) and IIRS. The main function of the Centre is to promote an increased and more effective use of microelectronics by industries in Ireland.

The initial State financing of the MAC was in the form of a long-term loan from SFADCo, training grants from the IDA and current contributions from SFADCo and NBST. The Centre's operations are financed principally by fee income from clients supplemented by SFADCo and it is intended that it will depend totally on clients' fees for funding from 1987 onwards. The Centre had a turnover of Ir£431 000 in 1985 and employed 13 permanent staff.

Activities are concentrated on the applications of microelectronics in new products and automated processes, with ancillary services such as equipment rental and general technical consultancy. The activities extend from specifications, through design, component sourcing, prototype building, tooling, technical literature and sub-contractor sourcing to field trials for production start-up.

The resources of the Centre include a team of experienced design engineers and technicians, with a comprehensive range of electronic development equipment.

Board Iascaigh Mhara (BIM)

The main functions of the BIM are to promote or engage in any business conducive to the development of the sea fishing industry. Its main activities are in the areas of market and fisheries development, promotion of the fisheries industry and the provision and/or facilitation of financial support for the industry.

As part of its activities the BIM provides a product and process development service for the fish processing industry. New products are appraised and processes developed for companies with the objective of achieving as high an added value as possible through expansion and improvement of processing technology. Ir£18 000 was spent on this activity in 1985 -- less than 1 per cent of the Board's total expenditure.

Industrial Development Authority (IDA)

The Authority has national responsibility for the furtherance of industrial development in Ireland. It carries out promotional and publicity programmes at home and abroad; provides grants and other financial facilities for new and existing manufacturing and technical service industries; provides training grants towards the costs of training workers; constructs and administers industrial estates; acquires industrial sites and constructs advance factories or custom-built factories for approved projects at regional locations; promotes joint ventures and licensing agreements; provides financial and advisory services to meet the special needs of small industries and craft industries; undertakes national and regional industrial planning; evaluates the implications for industrial development in Ireland of EEC proposals and policies. The Authority employs 700 persons. Its activities include regional planning of industry, the construction, provision and management of industrial buildings and estates; the provision of grants and other financial facilities for industrial development, feasibility studies, product and process development and national and international promotion of Ireland as an industrial base.

The IDA supports the technological development of firms directly through the provision of feasibility and product and process development grants. It also provides grants to firms for technology acquisition through licensing and/or joint ventures. New product opportunities are also sought by IDA overseas staff with a view of arranging licensing or joint-venture agreements with suitable indigenous firms.

Expenditure on direct technology-related activities accounts for 5 per cent of the total IDA budget.

Shannon Free Airport Development Company (SFADCo)

SFADCo's main functions are to further the industrial and commercial development of the Shannon Free Airport zone, generally on an exclusive agency basis, and to further the development of small, indigenous industrial enterprises in the remainder of the Southwest region.

SFADCo is financed by grant-in-aid, by EEC grants, by rent and other income. Payments in support of technology in industry amounted to Ir£1.7 million in 1985 -- 18 per cent of total activity. In all 199 people are employed, 14 of whom are engaged in science and technology activities.

The main S&T support which SFADCo gives to industry is in the form of product and process grants which are available to small firms and the payments for which are reimbursed by the IDA. SFADCo also aids Shannon industry with product and process grants. Feasibility grants are also paid and these enable individuals, groups and firms seek out and evaluate prospective new product ventures, including licensing, and plant costings.

SFADCo also supports technology transfer by providing grants for the buying and selling of technology through licensing and joint ventures.

National Board for Science and Technology (NBST)

The NBST's main function is to further scientific and technological development in line with national economic and industrial needs.

The general functions of the NBST are to advise the Government and the Minister for Industry and Commerce on policy and planning for science and technology, including the deployment of public funds and the co-ordination of the activities of institutions in receipt of public monies for scientific and technological activities; it also has powers to promote and pilot research and other activities in science and technology.

The NBST administers a number of programmes which support the technological development of industry. A prime focus for such support is the fostering of links between researchers in the higher education sector and industries' need for R&D.

Another main focus is forging links to European co-operative programmes. The NBST represents the Minister for Industry and Commerce both bilaterally and with such organisations as the EEC Commission and Council and the European Space Agency. It also aims to secure EEC- and ESA-funded research, development and demonstration projects for Irish industry. Projects which are relevant to national economic needs and which help transfer technology to Irish participants are actively sought on behalf of Irish industry.

The NBST employs a total staff of 74, and received a grant-in-aid of Ir£3.6 million from the Department of Industry and Commerce in 1986.

Udaras Na Gaeltachta

Udaras Na Gaeltachta was established under the Udaras Na Gaeltachta Act, 1979. The objectives of An t-Udaras are as follows: to encourage the preservation and extension of the Irish language in the Gealtacht; to attract suitable native and foreign manufacturing projects to the Gaeltacht; to establish, develop and manage productive employment enterprises in the Gaeltacht; to participate in industries as an equity partner; and to provide services to assist new industries becoming established.

Udaras Na Gaeltachta is financed by a grant-in-aid, rents, the European Social Fund, repayable advances and other income.

Udaras Na Gaeltachta employs 147 people (plus 8 seconded from other organisations).

The main technological support which An t-Udaras gives to firms is in the form of R&D grants and feasibility grants which enable individuals, groups and firms to seek out and evaluate prospective new product ventures.

Irish Goods Council (IGC)

The Irish Goods Council employs 27 permanent staff and receives a grant-in-aid in the order of Ir£1 million. The general functions of the Irish Goods Council include responsibility for the market development and promotion of Irish products on the home market, and the provision of a range of marketing support services to Irish industry.

The IGC also operates a scheme in conjunction with the Youth Employment Agency for the placement of marketing graduates in small firms which do not already employ professional marketing personnel. The firms are paid grants towards the salaries of those placed.

National Enterprise Agency (NEA)

The National Enterprise Agency was established in 1981 as a limited liability company. It was financed by grants-in-aid for current and capital expenditures; it also received management fee income from its investments, which were expected to amount to Irf25 000 in 1985. Its activities have recently been subsumed into the National Development Corporation (NDC).

The general function of the NEA/NDC is to assist in the development of commercially viable business enterprise in Ireland, by selective investment of risk capital in projects with a potential for increasing productive employment. The NEA employed 12 staff in 1985, none of whom were permanent.

National Software Centre (NCS)

The National Software Centre was established in 1983 as a private company. It is an independent subsidiary of the IDA and its board of directors includes representatives of Irish and international business, educational institutes, the IDA and the NBST.

The aims of the NSC are to increase the technical capability of Irish software companies and to improve the international image of Ireland as a location for software development. Activities of the Centre include the marketing of services and market research, product development, research and development and advanced training courses. Consultancy will be carried out on a fee-paying basis for clients.

The NSC, which currently employs 16 staff, was initially financed by capital grants and equity from the IDA and it is intended to work progressively towards financial self-sufficiency as fee income increases.

1985 SCIENCE AND TECHNOLOGY ALLOCATIONS (1)

	Total capital allocation	Total current allocation	Expected non-exchequer (2)	Increase in total current allocation over 1984 outturn (%)
		Ir£ thousand		
Department of An Taoiseach				
Sectoral Development Committee	-	80	79	1.3
Natural History Museum	-	280	-	13.3
National Economic and Social Council	-	210	10	-12.9
Central Statistics Office				
Department of Finance				
Economic and Social Research Institute	-	1 500	423	5.7
Office of Public Works	320	167	-	9.2
State Laboratory	-	1 630	30	33.6
Valuation and Ordnance Survey	-	5 575	1 084	10.9
Department of the Public Service	-	1 390	-	12.8
Department of Justice	-	4 007	38	12.5
Department of the Environment	-	18	-	0.0
Medical Bureau of Road Safety	-	422	-	16.3
An Foras Forbartha	-	5 232	1 463	14.7
Department of Education	7 118	54 450	9 533	9.8
Higher Education Authority	8 610	76 279	22 775	8.7
Dublin Institute for Advanced Studies	-	1 141	-	1.0

	Total capital allocation	Total current allocation	Expected non-exchequer (2)	Increase in total current allocation over 1984 outturn
		Irf thousand		(%)
Department of Fisheries & Forestry				
Board Iascaigh Mhara	258	565	5	23.9
Central and Regional Fisheries Board	-	2 021	2	4.3
Salmon Research Trust	-	226	149	38.7
Roinn na Gealtachta				
Udaras na Gaeltachta	420	377	-	(3) -6.1
Department of Agriculture	1 014	18 335	1 514	-2.1
An Foras Taluntais	-	23 412	7 048	3.1
An Chomhairle Oiliuna Talmhaiochta	-	14 560	4 198	0.3
Department of Labour	-	63	-	-13.6
Youth Employment Agency	-	76	-	-42.4
Department of Industry, Trade, Commerce and Tourism	-	1 371	-	17.9
Institute for Industrial Research and Standards	200	13 930	5 932	5.5
Kilkenny Design Workshop	-	1 125	596	-0.4
Industrial Development Authority	21 621	815	1 244	(3) 25.7
Shannon Free Airport Development Company	1 191	460	-	(3) 27.0
Innovation Centre	-	600	104	0.7
National Microelectronics Research Centre	422	713	213	-4.3
Irish Productivity Centre	-	1 209	823	13.1
National Board for Science and Technology	-	2 677	-	8.9
Irish Goods Council	-	213	-	58.1
National Enterprise Agency	-	200	-	100.0
Patents Office	-	1 643	2 696	7.2
National Microelectronics Application Centre	-	431	288	-28.9
Department of Energy	-	712	-	28.5
Geological Survey of Ireland	-	1 503	27	17.8
Nuclear Energy Board	-	316	-	3.7

	Total capital allocation	Total current allocation	Expected non-exchequer (2)	Increase in total current allocation over 1984 outturn (%)
	Ir£ thousand			
Department of Communications				
Meteorological Service	287	6 227	-	7.1
Telecom Eireann	-	3 000	-	50.0
Department of Social Welfare	-	1 820	-	137.9
Department of Health	-	58 983	8 566	2.3
Medical Research Council	-	1 330	20	2.7
Medico-Social Research Board	-	1 081	21	15.2
Total	21 461	322 129	69 145	8.2

1.   When programmes funded by one organisation are performed by another, the funding is included in the latter organisation's allocation(s).

2.   Expected non-exchequer income is exclusively part of total current allocation with the exception of the Industrial Development Authority and Shannon Airport Development Company.

3.   Total allocation compared with total 1984 outturn prices.

OECD SALES AGENTS
DÉPOSITAIRES DES PUBLICATIONS DE L'OCDE

ARGENTINA - ARGENTINE
Carlos Hirsch S.R.L.,
Florida 165, 4° Piso,
(Galeria Guemes) 1333 Buenos Aires
Tel. 33.1787.2391 y 30.7122

AUSTRALIA-AUSTRALIE
D.A. Book (Aust.) Pty. Ltd.
11-13 Station Street (P.O. Box 163)
Mitcham, Vic. 3132   Tel. (03) 873 4411

AUSTRIA - AUTRICHE
OECD Publications and Information Centre,
4 Simrockstrasse,
5300 Bonn (Germany)   Tel. (0228) 21.60.45
Local Agent:
Gerold & Co., Graben 31, Wien 1 Tel. 52.22.35

BELGIUM - BELGIQUE
Jean de Lannoy, Service Publications OCDE,
avenue du Roi 202
B-1060 Bruxelles   Tel. (02) 538.51.69

CANADA
Renouf Publishing Company Ltd/
Éditions Renouf Ltée,
1294 Algoma Road, Ottawa, Ont. K1B 3W8
Tel: (613) 741-4333
Toll Free/Sans Frais:
Ontario, Quebec, Maritimes:
1-800-267-1805
Western Canada, Newfoundland:
1-800-267-1826
Stores/Magasins:
61 rue Sparks St., Ottawa, Ont. K1P 5A6
Tel: (613) 238-8985
211 rue Yonge St., Toronto, Ont. M5B 1M4
Tel: (416) 363-3171
Sales Office/Bureau des Ventes:
7575 Trans Canada Hwy, Suite 305,
St. Laurent, Quebec H4T 1V6
Tel: (514) 335-9274

DENMARK - DANEMARK
Munksgaard Export and Subscription Service
35, Nørre Søgade, DK-1370 København K
Tel. +45.1.12.85.70

FINLAND - FINLANDE
Akateeminen Kirjakauppa,
Keskuskatu 1, 00100 Helsinki 10   Tel. 0.12141

FRANCE
OCDE/OECD
Mail Orders/Commandes par correspondance :
2, rue André-Pascal,
75775 Paris Cedex 16
Tel. (1) 45.24.82.00
Bookshop/Librairie : 33, rue Octave-Feuillet
75016 Paris
Tel. (1) 45.24.81.67 et/ou (1) 45.24.81.81
Principal correspondant :
Librairie de l'Université,
12a, rue Nazareth,
13602 Aix-en-Provence   Tel. 42.26.18.08

GERMANY - ALLEMAGNE
OECD Publications and Information Centre,
4 Simrockstrasse,
5300 Bonn   Tel. (0228) 21.60.45

GREECE - GRÈCE
Librairie Kauffmann,
28, rue du Stade, 105 64 Athens Tel. 322.21.60

HONG KONG
Government Information Services,
Publications (Sales) Office,
Beaconsfield House, 4/F.,
Queen's Road Central

ICELAND - ISLANDE
Snæbjörn Jónsson & Co., h.f.,
Hafnarstræti 4 & 9,
P.O.B. 1131 – Reykjavik
Tel. 13133/14281/11936

INDIA - INDE
Oxford Book and Stationery Co.,
Scindia House, New Delhi 1   Tel. 45896
17 Park St., Calcutta 700016   Tel. 240832

INDONESIA - INDONÉSIE
Pdii-Lipi, P.O. Box 3065/JKT.Jakarta
Tel. 583467

IRELAND - IRLANDE
TDC Publishers - Library Suppliers,
12 North Frederick Street, Dublin 1.
Tel. 744835-749677

ITALY - ITALIE
Libreria Commissionaria Sansoni,
Via Lamarmora 45, 50121 Firenze
Tel. 579751/584468
Via Bartolini 29, 20155 Milano   Tel. 365083
Sub-depositari :
Editrice e Libreria Herder,
Piazza Montecitorio 120, 00186 Roma
Tel. 6794628
Libreria Hœpli,
Via Hœpli 5, 20121 Milano   Tel. 865446
Libreria Scientifica
Dott. Lucio de Biasio "Aeiou"
Via Meravigli 16, 20123 Milano   Tel. 807679
Libreria Lattes,
Via Garibaldi 3, 10122 Torino   Tel. 519274
La diffusione delle edizioni OCSE è inoltre
assicurata dalle migliori librerie nelle città più
importanti.

JAPAN - JAPON
OECD Publications and Information Centre,
Landic Akasaka Bldg., 2-3-4 Akasaka,
Minato-ku, Tokyo 107   Tel. 586.2016

KOREA - CORÉE
Kyobo Book Centre Co. Ltd.
P.O.Box: Kwang Hwa Moon 1658,
Seoul   Tel. (REP) 730.78.91

LEBANON - LIBAN
Documenta Scientifica/Redico,
Edison Building, Bliss St.,
P.O.B. 5641, Beirut   Tel. 354429-344425

MALAYSIA - MALAISIE
University of Malaya Co-operative Bookshop
Ltd.,
P.O.Box 1127, Jalan Pantai Baru,
Kuala Lumpur   Tel. 577701/577072

NETHERLANDS - PAYS-BAS
Staatsuitgeverij
Chr. Plantijnstraat, 2 Postbus 20014
2500 EA S-Gravenhage   Tel. 070-789911
Voor bestellingen: Tel. 070-789880

NEW ZEALAND - NOUVELLE-ZÉLANDE
Government Printing Office Bookshops:
Auckland: Retail Bookshop, 25 Rutland Street,
Mail Orders, 85 Beach Road
Private Bag C.P.O.
Hamilton: Retail: Ward Street,
Mail Orders, P.O. Box 857
Wellington: Retail, Mulgrave Street, (Head
Office)
Cubacade World Trade Centre,
Mail Orders, Private Bag
Christchurch: Retail, 159 Hereford Street,
Mail Orders, Private Bag
Dunedin: Retail, Princes Street,
Mail Orders, P.O. Box 1104

NORWAY - NORVÈGE
Tanum-Karl Johan
Karl Johans gate 43, Oslo 1
PB 1177 Sentrum, 0107 Oslo 1Tel. (02) 42.93.10

PAKISTAN
Mirza Book Agency
65 Shahrah Quaid-E-Azam, Lahore 3 Tel. 66839

PORTUGAL
Livraria Portugal,
Rua do Carmo 70-74, 1117 Lisboa Codex.
Tel. 360582/3

SINGAPORE - SINGAPOUR
Information Publications Pte Ltd
Pei-Fu Industrial Building,
24 New Industrial Road No. 02-06
Singapore 1953   Tel. 2831786, 2831798

SPAIN - ESPAGNE
Mundi-Prensa Libros, S.A.,
Castelló 37, Apartado 1223, Madrid-28001
Tel. 431.33.99
Libreria Bosch, Ronda Universidad 11,
Barcelona 7   Tel. 317.53.08/317.53.58

SWEDEN - SUÈDE
AB CE Fritzes Kungl. Hovbokhandel,
Box 16356, S 103 27 STH,
Regeringsgatan 12,
DS Stockholm   Tel. (08) 23.89.00
Subscription Agency/Abonnements:
Wennergren-Williams AB,
Box 30004, S104 25 Stockholm.
Tel. (08)54.12.00

SWITZERLAND - SUISSE
OECD Publications and Information Centre,
4 Simrockstrasse,
5300 Bonn (Germany)   Tel. (0228) 21.60.45
Local Agent:
Librairie Payot,
6 rue Grenus, 1211 Genève 11
Tel. (022) 31.89.50

TAIWAN - FORMOSE
Good Faith Worldwide Int'l Co., Ltd.
9th floor, No. 118, Sec.2
Chung Hsiao E. Road
Taipei   Tel. 391.7396/391.7397

THAILAND - THAILANDE
Suksit Siam Co., Ltd.,
1715 Rama IV Rd.,
Samyam Bangkok 5   Tel. 2511630

TURKEY - TURQUIE
Kültur Yayinlari Is-Türk Ltd. Sti.
Atatürk Bulvari No: 191/Kat. 21
Kavaklidere/Ankara   Tel. 25.07.60
Dolmabahce Cad. No: 29
Besiktas/Istanbul   Tel. 160.71.88

UNITED KINGDOM - ROYAUME-UNI
H.M. Stationery Office,
Postal orders only:
P.O.B. 276, London SW8 5DT
Telephone orders: (01) 622.3316, or
Personal callers:
49 High Holborn, London WC1V 6HB
Branches at: Belfast, Birmingham,
Bristol, Edinburgh, Manchester

UNITED STATES - ÉTATS-UNIS
OECD Publications and Information Centre,
Suite 1207, 1750 Pennsylvania Ave., N.W.,
Washington, D.C. 20006 - 4582
Tel. (202) 724.1857

VENEZUELA
Libreria del Este,
Avda F. Miranda 52, Aptdo. 60337,
Edificio Galipan, Caracas 106
Tel. 32.23.01/33.26.04/31.58.38

YUGOSLAVIA - YOUGOSLAVIE
Jugoslovenska Knjiga, Knez Mihajlova 2,
P.O.B. 36, Beograd   Tel. 621.992

Orders and inquiries from countries where Sales
Agents have not yet been appointed should be sent
to:
OECD, Publications Service, Sales and
Distribution Division, 2, rue André-Pascal, 75775
PARIS 16.

Les commandes provenant de pays où l'OCDE n'a
pas encore désigné de dépositaire peuvent être
adressées à :
OCDE, Service des Publications. Division des
Ventes et Distribution. 2. rue André-Pascal. 75775
PARIS CEDEX 16.

70431-01-1987

OECD PUBLICATIONS, 2, rue André-Pascal, 75775 PARIS CEDEX 16 - No. 43915 1987
PRINTED IN FRANCE
(92 87 01 1) ISBN 92-64-12918-9